Simulation in Chassis Technology

Dirk Adamski

Simulation in Chassis Technology

A Practice-oriented Introduction
to the Creation of Component
and Full Vehicle Models Using
the Method of Multi-Body Systems

 Springer

Dirk Adamski
Automotive and Aerospace Engineering
UAS Hamburg
Hamburg, Germany

ISBN 978-3-658-30677-9   ISBN 978-3-658-30678-6   (eBook)
https://doi.org/10.1007/978-3-658-30678-6

The translation was done with the help of artificial intelligence (machine translation by the service DeepL.com). A subsequent human revision was done primarily in terms of content.

This Springer imprint is published by the registered company Springer Fachmedien Wiesbaden GmbH, part of Springer Nature.
The registered company address is: Abraham-Lincoln-Str. 46, 65189 Wiesbaden, Germany

Writing is easy. All you have to do is cross out the wrong words

Mark Twain

Preface

The method of multi-body systems (MBS) described in this book is a standard for the vehicle dynamics and ride comfort simulation of chassis and full vehicles. Here, however, it is not the method itself that is in the foreground, but its application. Above all, the implementation of questions which a computational engineer in the abovementioned disciplines is confronted with on a daily basis will be discussed here. This means that the formulas and derivations necessary for understanding the method are usually not to be found here. However, you will find many practical examples and of course the equations used for this. Of course, the profession of computational engineer and the fear of differential equations do not fit together. Nevertheless, I assume, it is enough for many people that the formalisms of the MBS method have already been processed in simulation programs and that they are familiar with the *essential* I want to dedicate to everyday business. The work order usually means calculating a tuning variant or developing the model of an active system rather than programming a new simulation tool. However, it is more about the description of the models of chassis components than about the operation of a commercial simulation tool.

This book is aimed primarily at developers of passenger car chassis, but the majority can nevertheless also be used for the development of commercial vehicle chassis. In most cases, the described models are held sufficiently general to be suitable for larger and heavier vehicles, especially through their parameterization—the physics are the same. However, it must always be considered whether the approach of multi-body systems with ideally rigid components used here can also be used for these questions. At least one hybrid simulation with FE components should be aimed at. In the case of two-wheeled vehicles, the structure of the chassis is different for some parts due to the structure of the vehicle, and the same for others—this also results in a certain intersection here.

I will not go into the basic functions of the individual chassis components or only rudimentarily. I assume that everyone who dares to approach the topic of simulation in the chassis has understood these basics and knows the appropriate standard literature.

The target group of this book is primarily engineers. But I would ask you to define this term somewhat more broadly than just considering the graduates of engineering studies. I am well aware that many mathematicians and physicists work in this field. But in my opinion, they then work as engineers, regardless of their academic qualifications.

I am not afraid to remain on the surface at some points and to convey more the operational and less the theoretical content. Fortunately, there are shelf meters of theoretical literature—at least for the first part of this book. My ambition was not to add a few more inches there. Anyone who is not satisfied with the explanations after the first reading, because he wants to, should, or must dive deeper into the area, is recommended this only as a suggestion for further studies. Those who stand helplessly in front of the meters of shelves should take their first steps to show them the way to more.

As a professor, I cannot help but remind you of the basic subjects you have learned about in your studies, some of which may have fallen into oblivion. We will need it in one place or another.

The topic of simulation has accompanied me since my own studies. Even as a student assistant at the mechatronics department of the former Gerhard Mercator University in Duisburg, Germany, I used the vehicle dynamics simulation package *FASIM_C++*, developed by them [AdSH97], the further development of which I later did my doctorate on [Adam01]. There, I mainly got to know the software side of simulation. In the first years of my professional career at Daimler AG, the development of the electrohydraulic brake system SBC was accompanied by hardware-in-the-loop simulations. During my time as a computational engineer at the same company, I developed and used large, complex full vehicle models for ride comfort simulation using a commercial simulation environment. I experienced the introduction phase of the so-called digital prototypes, which are standard in automotive development today, in several projects [AdJD07]. Today, as professor for *Testing and Simulation in Chassis Technology* at the Hamburg University of Applied Sciences, I teach this subject there.

Hamburg, Germany   Dirk Adamski
March 2020

References

[Adam01] ADAMSKI, D.: *Komponentenbasierte Simulation mechatronischer Systeme*, VDI-Fortschrittberichte Nr. 682, Reihe 10, VDI-Verlag, Duesseldorf, 2001

[AdJD07] ADAMSKI, D., JUST, W. AND DRAGON, L.: Subjektive Bewertung des Ride Komforts von Digitalen Prototypen mit Hilfe eines Ride Simulators, in *Erprobung und Simulation in der Fahrzeugentwicklung*, VDI-Berichte 1990, S.289-302, Wuerzburg, 2007

[AdSH97] ADAMSKI, D., SCHUSTER, CH. AND HILLER, M.: Fahrdynamiksimulation mit FASIM_C++ als Beispiel für die Modellierung mechatronischer Systeme, in *Mechatronik im Maschinen- und Fahrzeugbau*, VDI-Bericht 1315, 117-141, VDI-Verlag, Moers, 1997

Contents

Part I Introduction to Simulation

1 Simulation Methods 3
   1.1   What Is Simulation? 3
   1.2   Approaches 4
          1.2.1   Finite Element Analysis (FEA) 4
          1.2.2   Multi-body Systems (MBS) 5
          1.2.3   Block-Oriented Methods 10
   References 11

2 Systems Engineering 13
   2.1   Concept of Systems 13
          2.1.1   System Boundary 14
          2.1.2   Causality 16
          2.1.3   Transmission Behavior 17
          2.1.4   Range of Values 18
          2.1.5   Linear and Non-linear Systems 19
   2.2   System Behavior 21
          2.2.1   Systems with and Without Memory 21
          2.2.2   Change Behavior 22
   2.3   Questions from the Given System Structure 23
          2.3.1   System Analysis 23
          2.3.2   System Identification 24
          2.3.3   System Control 25
   References 26

3 Modeling 27
   3.1   At the Beginning, There Is the Problem 27
   3.2   The Difference Between Erroneous and False 29
   3.3   Methods for Modeling 30
          3.3.1   Induction 30
          3.3.2   Deduction 31

	3.3.3	Method of Choice	31
3.4	Model Classes		32
	3.4.1	Physical Models	32
	3.4.2	Behavioral Models	32
3.5	Problem Analysis		34
	3.5.1	Analysis of the Question	34
	3.5.2	Analysis of the System	34
3.6	Model Design		35
	3.6.1	Simulation Method	35
	3.6.2	Implementation of the Problem Analysis	35
3.7	Verification		36
3.8	Validation		36
	3.8.1	Basic Procedure	36
	3.8.2	Comparison of Measurement and Simulation	38
	3.8.3	Comparison of Simulation and Simulation	41
	3.8.4	Validation with Full Vehicle Measurements	41
3.9	Single or Multiple Use		42
	3.9.1	Modularized or Monolithic?	42
	3.9.2	Separation of Data and Model	44
References			44

4	**Numerical Analysis: The Problem with the Beginning**		**45**
4.1	Who Is EULER?		45
4.2	Initial Value Problems or Numerical Integration of Differential Equations		46
	4.2.1	The Initial Value Problem	46
	4.2.2	Numerical Integration	47
4.3	Numerical Integration of First Order Differential Equations		47
	4.3.1	A Simple Example	48
	4.3.2	Polygonal Method According to EULER	49
	4.3.3	Types of Errors	50
	4.3.4	Convergence and Stability	53
4.4	Integration Methods		57
	4.4.1	Method Overview	57
	4.4.2	Implicit EULER's Method	58
	4.4.3	RUNGE-KUTTA Method	60
	4.4.4	ADAMS Method	61
	4.4.5	BDF Method	62
4.5	Interpolation and Extrapolation Methods		63
	4.5.1	Interpolation	63
	4.5.2	Extrapolation	65

	4.6	Functions for Fading in and Fading out	66
		4.6.1 Linear	66
		4.6.2 Exponential	67
		4.6.3 Trigonometric	70
	References		72
5	**Simulation Tools**		**73**
	5.1	Tool Selection	73
		5.1.1 In-House Solution	73
		5.1.2 Commercial Product	74
	5.2	Basic Structure of a Simulation Environment	75
		5.2.1 Preprocessor	75
		5.2.2 Solver	77
		5.2.3 Postprocessor	78
	5.3	Interfaces for Co-simulation	82
		5.3.1 Controller Import	83
		5.3.2 Importing MBS Models	83
		5.3.3 Online Simulation	85
		5.3.4 Potential Communication Problems	85
	References		87
6	**Simulation Process**		**89**
	6.1	Parameter Procurement	90
		6.1.1 Need of Parameters	90
		6.1.2 Naming of Parameters	91
		6.1.3 Unit-Related Parameters	94
		6.1.4 One-Dimensional Parameters	94
		6.1.5 Multidimensional Parameters	95
		6.1.6 Vehicle Reference System	100
		6.1.7 Mass Properties	102
	6.2	Pre-Simulation Phase	104
		6.2.1 Consistency of Data and Model	104
		6.2.2 Model Diversity	105
		6.2.3 Simulation History	105
	6.3	Simulation Phase	106
		6.3.1 Local or Distributed	106
		6.3.2 Copying Procedure	106
		6.3.3 Licenses	107
	6.4	Post-Simulation Phase	107
		6.4.1 Documentation of the Simulation	107
		6.4.2 Archiving	107

        6.4.3   Motivation for Documentation   108
  6.5   Reproducibility of the Simulation Results   108
  References   110

Part II   Simulation in Chassis Technology

7   Modeling of Chassis Components   113
  7.1   Fields of Application and Limits of Simulation   113
        7.1.1   Vehicle Dynamics and Driver Assistance Systems   114
        7.1.2   Ride Comfort   114
        7.1.3   Load Data Prediction   116
        7.1.4   Use of Simulators   116
        7.1.5   Potential of Calculation or Undiscovered Treasures   117
  7.2   Complexity of Models   118
        7.2.1   Maintenance and Modifications   118
        7.2.2   Computing Time Requirement   119
        7.2.3   Parameter Requirements   119
  7.3   Simple Model Approaches   120
  7.4   Where Is the Right Information?   120
  7.5   Planning and Evaluation of Maneuvers   121
        7.5.1   Settling Time   121
        7.5.2   Length and Duration of the Maneuver   122
  References   123

8   Kinematics and Compliance   125
  8.1   Modeling of the Kinematics   125
        8.1.1   Mechanism-Oriented Models   125
        8.1.2   Map-Oriented Models   133
        8.1.3   Behavior-Oriented Models   134
  8.2   Modeling of the Compliance   134
        8.2.1   Elastic Chassis Parts   134
        8.2.2   Secondary Spring Rate   137
  8.3   Simple Elastomeric Bearing Models   137
        8.3.1   Linear Parameterization   139
        8.3.2   Non-linear Parameterization   145
        8.3.3   Influence of Amplitude and Frequency of Excitation   148
  8.4   Basics of Typical Elastomer Bearing Models   149
        8.4.1   MAXWELL Element   149
        8.4.2   KELVIN-VOIGT Element   150
        8.4.3   Combination of Several Elements   152
  8.5   Special Chassis Bearings   152
        8.5.1   Hydromounts   152

	8.5.2	Top Mounts	154
8.6	Adjustment of Kinematics and Compliance with Measurements		156
	8.6.1	Creation of Wheel Travel Curves	156
	8.6.2	Deviations in Vehicle Level	157
	8.6.3	Deviations in Kinematics or Compliance	157
	8.6.4	Additional Springs	157
	8.6.5	Suspension Spring Stiffness	158
	8.6.6	Anti-roll Bar Stiffness	159
References			159

9 Springs 161

9.1	Steel Springs	162
	9.1.1	Coil Spring	162
	9.1.2	Leaf Spring	165
	9.1.3	Torsion Bar	170
	9.1.4	Anti-roll Bar	171
9.2	Air Spring	173
	9.2.1	Determination of Quasi-static Stiffness	173
	9.2.2	Determination of Dynamic Stiffness	174
	9.2.3	Use of Measured Characteristic Curves	176
	9.2.4	Level Control	177
9.3	Bound and Rebound Bump Stops	177
	9.3.1	Bound Bump Stop	177
	9.3.2	Rebound Bump Stop	178
	9.3.3	Combination	179
9.4	Spring Ratio	179
References			181

10 Damping and Friction 183

10.1	Dampers	183
	10.1.1	Force Law and Damper Characteristic Curve	183
	10.1.2	Kinematics and Mass	187
	10.1.3	Damper Ratio	187
	10.1.4	Gas Spring Forces	188
	10.1.5	Seals and Friction	188
	10.1.6	Temperature Influence	188
	10.1.7	Complex Damper Models	189
10.2	Friction	189
	10.2.1	Coulomb's Friction	190
	10.2.2	Fictitious Total Friction	192
References			192

11 Steering 195
   11.1   Simple Steering Models 196
   11.2   Steering Train 199
       11.2.1   Steering Gear 199
       11.2.2   Steering Column 200
       11.2.3   Steering Wheel 202
   11.3   Power Steering 203
       11.3.1   Hydraulic Power Steering (HPS) 203
       11.3.2   Electrohydraulic Power Steering (EHPS) 204
       11.3.3   Electric Power Steering (EPS) 205
   References 207

12 Tires and Roads 209
   12.1   General Requirements for Tire Models 212
       12.1.1   Modeling the Contact Patch 212
       12.1.2   Friction Contact and Slip Definition 213
       12.1.3   Limits of the Slip Definition 217
       12.1.4   Standard Tyre Interface 218
   12.2   Tire Models for Vehicle Dynamics 219
       12.2.1   Magic Formula 220
       12.2.2   MF-Tyre and MF-SWIFT 220
       12.2.3   HSRI-Model 222
   12.3   Tire Models for Ride Comfort and Load Prediction 223
       12.3.1   FTire 224
       12.3.2   RMOD-K 224
       12.3.3   CDTire 225
   12.4   Parameterization of the Tire Models 226
       12.4.1   Parameterization Process 226
       12.4.2   Measurement of Tire Parameters 228
       12.4.3   Models of Varying Complexity 229
   12.5   Modeling the Road 229
       12.5.1   Measurement Methods for Road Profiles 229
       12.5.2   Topology of the Road 231
       12.5.3   Single Events 233
       12.5.4   Periodic Excitation 235
       12.5.5   Stochastic Excitations 235
   References 236

13 Drive Train 239
   13.1   Specification of the Drive Torque 239
   13.2   Engine and Gearbox 240
       13.2.1   Engine Map and Time Response 240

13.2.2 Mass Data 241
13.2.3 Engine Mounts 242
13.3 Axle and Center Differentials 243

14 Brake System 245
14.1 Specification of the Brake Torque 245
14.2 Brake Circuits 247
14.3 Brake Force Proportioning 247
14.4 Functional Chain from Driver to Wheel Brake 248
14.5 Brake Torque at the Wheel Brake 252
14.5.1 Drum Brake 252
14.5.2 Disc Brake 253
14.6 Braking to a Standstill 254
14.7 Coefficient of Friction and Temperature Behavior 255
References 255

15 Vehicle Body 257
15.1 Body in White 257
15.1.1 Preparation of the FE Model 258
15.1.2 Modal Reduction 259
15.2 Total Mass 260
15.2.1 Mass Distribution 260
15.2.2 Use of One Correction Mass 261
15.2.3 Use of Several Correction Masses 262
15.2.4 Conclusion 262
15.3 Aerodynamics 262
15.3.1 Wind Resistance 263
15.3.2 Crosswind 263
15.3.3 Buoyancy 264
References 265

16 The Simulated Driver 267
16.1 Speed Control 268
16.1.1 Initial Value 268
16.1.2 Open-Loop Maneuvers 269
16.1.3 Closed-Loop Maneuver 270
16.2 Steer Control 272
16.2.1 Open-Loop Maneuver 272
16.2.2 Closed-Loop Maneuver 272
16.3 Complex Driver Models 275
References 275

17 The Vehicle Model as a Controlled System 277

    17.1   Development of Control Systems 277

          17.1.1   Software-in-the-Loop 278

          17.1.2   Hardware-in-the-Loop 279

    17.2   Sensors 282

    17.3   Actuators 282

    References 283

Index 285

List of Symbols

a	Magnitude of the translational acceleration
A	Cross-sectional area
b	Width
c	Spring stiffness, aerodynamic coefficients
c_m	Specific thermal capacity
c_W	Drag coefficient
C	Capacity
C^*	Internal brake ratio
d	Damping constant
e	Local error
$\vec{e}_x, \vec{e}_y, \vec{e}_z$	Unit vectors
E	Energy, modulus of elasticity
f	Number of degrees of freedom, frequency
f_G	Joint degree of freedom
F	Force
g	Gravity
G	Shear modulus
h	Integration step size, height
i	Current, ratio
I	Moment of inertia of area
k	General coefficient
l	Length
L	Inductance
m	Mass
M	Torque
n	Number, rotational speed
n_B	Number of bodies
n_J	Number of joints
n_L	Number of kinematical loops
p	Pressure, gain

Q	Thermal energy
r	Number of isolated degrees of freedom, general radius
R	Ohmic resistor, curve radius
R_L	Specific gas constant
s	Displacement
S	Path length
t	Time
T	Temperature
U	Internal energy
v	Magnitude of the translational velocity
V	Volume
w	Bending line
W	Work
x, y, z	Path coordinates in the Cartesian coordinate system
$\dot{x}, \dot{y}, \dot{z}$	Translational velocities
$\ddot{x}, \ddot{y}, \ddot{z}$	Translational accelerations
α	Heat transfer coefficient, general angle, slip angle
β	Side slip angle, width ratio, bend angle
γ	Torsion angle, camber angle
δ	Lost angle, steer angle
ε	Stretching
Θ	Moment of inertia
κ	Ratio of cross-sectional area, isentropic exponent
λ	Coefficient of slip
μ	Coefficient of friction
ρ	Density
σ	Normal stress
τ	Wind angle
$\ddot{\varphi}$	Rotational acceleration
Ψ	Yaw angle
ω	Rotational velocity
ω_0	Undamped natural frequency

List of Abbreviations

ASCII	American Standard Code for Information Interchange
ASAM	Association for Standardization of Automation and Measuring Systems
BDF	Backward differential functions
CAD	Computer-aided design
CAN	Controller area network
CFD	Computational fluid dynamics
CPU	Central processing unit
DAE	Differential algebraic equations
DASSL	Differential–algebraic system solver
DIN	Deutsches Institut für Normung e. V.
DLL	Dynamic-link library
EHPS	Electro-hydraulic power steering
EMC	Electromagnetic compatibility
EPS	Electric power steering
ESC	Electronic stability control
FE	Finite element
FEA	Finite element analysis
FEM	Finite element method
FFT	Fast Fourier transformation
GIGO	Garbage in–garbage out
GPS	Global positioning system
GRP	Glass-reinforced plastic
HiL	Hardware-in-the-loop
HPS	Hydraulic power steering
ISO	International Organization for Standardization
K&C	Kinematics and compliance
MBS	Multi-body systems
NURBS	Non-uniform rational B-spline
NVH	Noise vibration harshness
ODE	Ordinary differential equation

OEM   Original equipment manufacturer
PDE   Partial differential equation
PSD   Power spectral density
RBE   Rigid body element
SAE   Society of Automotive Engineers
SBC   Sensotronic Brake Control
SiL   Software-in-the-loop
TCS   Traction control system
TNO   Toegepast Natuurwetenschappelijk Onderzoek
VDA   Verband der Automobilindustrie e. V.
VDI   Verband Deutscher Ingenieure e. V.

Introduction to Simulation

The topic *Simulation in Chassis Technology* starting directly with the models of parts and components will overwhelm many newcomers to the subject. For this reason, this first part precedes the main topic in order to explain the most important questions for the modeling and simulation of chassis, starting with the basics. Even if engineers like to jump right into the problem, it is often helpful to take a step back and look at the whole thing. This is especially the case if the method is not only to be simulated once, but is also to be used regularly and productively.[1] Furthermore, the term simulation always refers to the computational simulation, i.e. calculations are carried out with the aid of computer technology and suitable model approaches.

---

[1] At several points, I will talk about the simulation's productive use. That means the simulation is used consistently and with effect on the product, meaning it is integrated into the development process. In other cases, simulation is used rather sporadically and primarily to gain knowledge.

Simulation Methods

1

> Reality is neither in the world nor in our models, but in the
> process of working back and forth between world and model.
>
> E.A. Singer in [Shan75]

1.1   What Is Simulation?

Let us start with the colloquial variant of simulation. "*He is only simulating*" is commonly used to describe a faked illness, the *spontaneous* for example, when it comes to avoid taking an imminent examination. This fits the actual meaning of the original Latin word quite well (*similar:* to pretend, to emulate). The Brockhaus[1] also begins its description with: "The conscious pretense of states, e.g. the conscious (mostly purpose-oriented) pretense of illnesses." [Broc14].

But in this book we want to follow the concept of reproduction rather than pretense. The Brockhaus goes on to say: "The model-like representation or reproduction of certain aspects of an existing or to be developed cybernetic system or process".

We will deal with the concept of the system in more detail in the following Chap. 2. This term is often used in the most diverse contexts, so that a limitation to the technical world is necessary—even if the systems engineering is universal and it is not limited to the technology. We will then deal with the modeling of systems in Chap. 3.

---

[1]In Germany a well-known encyclopedia—in times before Wikipedia.

© Springer Fachmedien Wiesbaden GmbH, part of Springer Nature 2021
D. Adamski, *Simulation in Chassis Technology*,
https://doi.org/10.1007/978-3-658-30678-6_1

1.2   Approaches

In order to represent a real system as a model in the simulation, there are different approaches. None of these methods is the only true and answering one. It is the different questions that justify the methods and in part also limit their possibilities.

This book takes a look at the world of models from the perspective of technical mechanics. The main focus is therefore on movement behavior (kinematics) and on the applying and occurring forces and torques (kinetics). In this world, systems are differentiated according to whether the system parameters are distributed (continuum) or concentrated (rigid bodies). In the first case, the system behavior must be described by Partial Differential Equations, since it depends on location and time. In the second case by Ordinary Differential Equations, since there is only a time dependence.

Most of us are familiar with the approach of concentrated parameters, since this is the main topic of the basic lectures of technical mechanics. Under the assumption of a homogeneous, isotropic mass distribution, this mass is concentrated in a single point— the center of gravity. Students of engineering sciences have been familiar with the approach of letting forces apply a body's center of gravity since their first semester. Also, all other forces apply by this idea at discrete points. Line and area loads are converted accordingly into a substitute force.

The approaches and tools typically used in chassis development are briefly presented below. If you would like to learn more about these methods, please make use of to the wide range of technical literature provided. At this point in text, we will leave it at an overview. Subsequently, I will essentially limit myself to the method of multi-body systems.

1.2.1   Finite Element Analysis (FEA)

The finite element method divides bodies or surfaces into a large but finite number of geometrically simple partial bodies or surfaces, the elements. They are then connected to each other at discrete points, the so-called nodes. Due to their simplicity (tetrahedron, triangles, etc.) the elements can be calculated quickly.[2] Their geometry is known and their deformation behavior can be stored as a material law so that distortions and stresses can be calculated. Since the local deformations are often of small order, the correlations can be linearized well. However, there are also applications in which linearization is no longer possible and nonlinear approaches must be used accordingly (e.g. crash simulation). There are respective solvers for both classes of applications.

---

[2]Of course, with time, more complex elements have been developed. However, that does not change the main idea about finite elements at all.

The method has been continuously developed since the 1940s, so that the complexity has increased immensely from initially simple beam and truss structures to today's very large full vehicle models [Klei12].

A major advantage of this method is that it is often possible to directly access the design data from the CAD system and thus represent a closed process chain. A manual transformation of the data is usually not necessary, since the FE programs often have corresponding interfaces to the CAD programs and can also transfer the material values and weight data. However, the type and degree of networking are still strongly dependent on the experience of the implementer and the question posed.

The main applications of this method in chassis development are the areas of fatigue, crash, vibration analysis, acoustics and compliance design. An extension represents the flow simulation (*Computational Fluid Dynamics*, CFD), which is used, for example, for the flow around the brake in order to be able to prove that it has sufficient cooling.

The models used for NVH analyses (*Noise Vibration Harshness*) are also applied in the frequency range and can be used even in acoustics. Stress analyses are carried out in the area of fatigue with which critical components or component locations are to be determined which could be prone to damage under certain circumstances. In this case, more and more use is being made of the MBS simulation of vehicles on digitized tracks [Adam07] or the calculation with test bench models [Neuw04] for load application in addition to the use of measurement data from real vehicles or components. In terms of lightweight construction, the analyses are also used to construct stress-optimized structures, i.e. only to use material where it is needed.

The full vehicle models, such as those required for NVH, crash or fatigue calculations, have become increasingly finely meshed in recent years, so that they already contain several million degrees of freedom. This trend seems to be unstoppable and only limited by the respective computing power. Due to these model sizes one is far away from real-time capability. Even highly simplified models are generally not suitable for real-time applications.[3] The role of these full vehicle models in the context of chassis simulation is discussed in more detail in Chap. 15.

1.2.2 Multi-body Systems (MBS)

With the method of multi-body systems, mechanisms are generally seen as connections of ideally rigid bodies, whose mass properties are concentrated in the center of gravity, which have no expansion of their own and have ideal joints. Furthermore, massless force elements, such as springs or dampers, and constraints, can act between the bodies. The number of degrees of freedom is smaller by orders of magnitude than with the FEA approaches. They range from single-digit values for simple models to a few hundred

---

[3]Those who have troubles understanding the term real-time, should read Chap. 17.

degrees of freedom for very complex full vehicle models. This allows completely different calculation times than for FE calculations, so that real-time capability can be guaranteed for some models, depending on the model complexity. In addition, large, non-linear movements can be calculated, such as those that occur when the wheel is deflected (see also Sect. 2.1.5). The calculations are usually made in the time domain. The basic idea of multi-body systems is immediately understandable for most engineers, since it corresponds with what they heard in their basic lectures on statics and dynamics within the field of technical mechanics in the first semesters of their studies.

The topology of the multi-body system has a large influence on the computing time. A distinction is made between an open (tree structure) and a closed (kinematic loop) topology (Fig. 1.1). For the tree structure, the number of joints n_J equals the number of bodies n_B and the number of degrees of freedom f.

$$f = n_B = n_J \tag{1.1}$$

However, for the kinematic loop, the number of joints is equal to the number of bodies plus the number of loops n_L.

$$n_J = n_B + n_L \tag{1.2}$$

The loop shown (plane four-bar linkage) has only one degree of freedom. All other joint angles are inevitably adjusted. Both topologies can be found in the chassis when the kinematics of the wheel suspension and the steering are modeled in detail. The question of this is necessary, will be dealt with in Chap. 8.

Complex kinematic loops, as they occur in many wheel suspension concepts, are usually no longer solvable closed. The solution can often only be found iteratively. This is where the computing time problem lies. In closed solvable systems the number of equations to be calculated is known a priori. If the solution has to be iterative, the number of function evaluations varies depending on the situation. The way in which the solution is achieved is different in the simulation programs available on the market and thus also the respective computing time requirement. The real-time condition is often by definition not achievable for such models.

With simulation tools for multi-body systems, a distinction must first be made between multi-purpose and special-purpose applications. The first group is not specifically designed for vehicle development, but is suitable for all mechanisms that can be described by this method (robots, machines, etc.). Due to the power of these tools, the training effort should not be underestimated. However, one is very flexible when it comes to implementing your own ideas and concepts later.

In comparison, the group of special purpose applications focuses on vehicle development and offers model catalogs and evaluation methods from this area. Some of the general programs can be adapted to vehicle industry-specific applications via special libraries. The

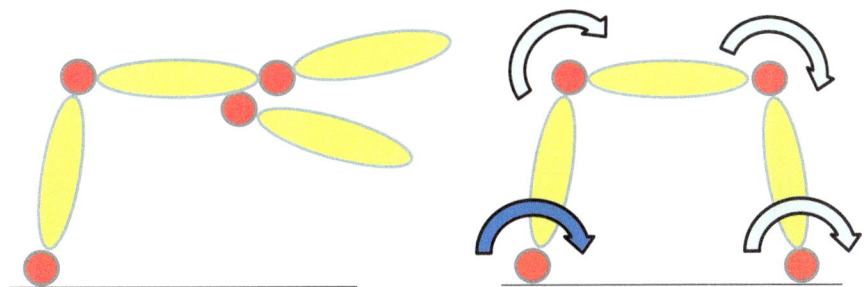

Fig. 1.1 Topology of a multi-body system (left open, right closed)

three main multi-purpose tools used internationally in the automotive industry all have such extensions.

The program Adams (*Automatic Dynamic Analysis of Mechanical Systems*) by the American company MSC Software is supported by the package Adams/car (Fig. 1.2). The German company Siemens PLM Software offers the simulation environment virtual.lab whose additional package vehicle motion (Fig. 1.3) contains the usual vehicle technical evaluations and libraries. Older computational engineers may still be familiar with the predecessor program DADS from this provider.

The German company Simpack AG has developed a program of the same name, Simpack, which is completed by the library Simpack Automotive+ (Fig. 1.4). Meanwhile it is part of SIMULIA Simulink of the French company Dassault Systèmes.

Examples of commercial special purpose tools are CarMaker from IPG Automotive (Fig. 1.5) or DYNA4 from Tesis Dynaware (Fig. 1.6). They can be used for vehicle dynamics simulation and are real-time capable. On the application side, the focus here is on the drive train and chassis, but there are also expansion modules, for example to carry out traffic simulations for driver assistance systems.

The simulation environments discussed here relieve the user of the necessity of knowing or having to understand the underlying formalisms for solving the equations of motion. In this way, the number of potential users is drastically increased. Nevertheless, their number is many times smaller than that of the much older finite element method. Attempts to couple the simulation environments with common CAD systems promise an increase in the number of users. This means that every designer can also carry out calculations that go beyond the previous possibilities of CAD systems. However, one should not underestimate the necessary know-how required for complex calculations. And certainly not that which is necessary for the interpretation of the results.

The few university chairs that deal with the further development of the method often have very efficient simulation tools at their disposal. Their large-scale commercial application is usually not possible due to a low user-friendliness and limited support possibilities. Understandably, the available resources are mostly invested in the development of algorithms.

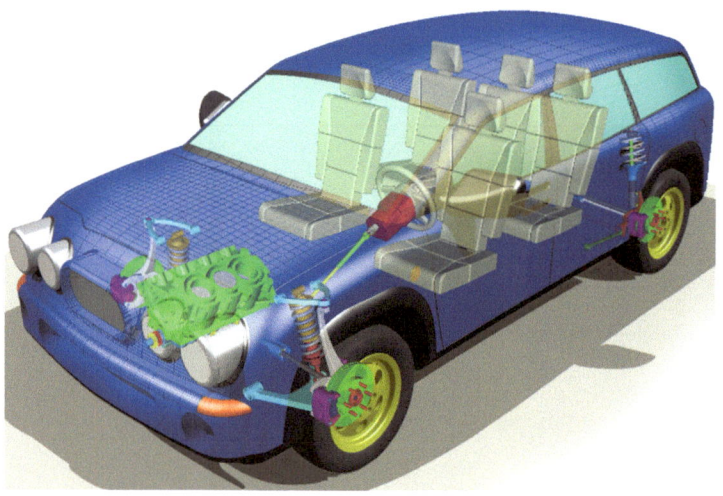

[MSC Software]

Fig. 1.2 Adams/car

[Siemens PLM Software]

Fig. 1.3 Vehicle motion

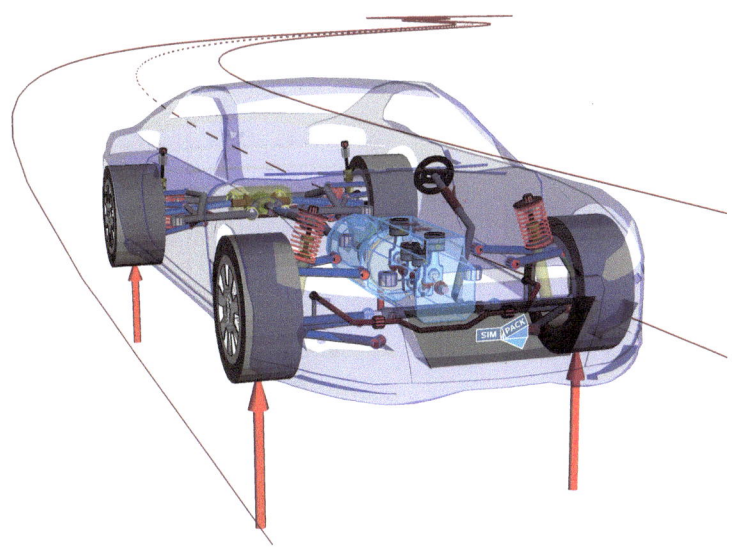

[Simpack AG]

Fig. 1.4 Simpack Automotive+

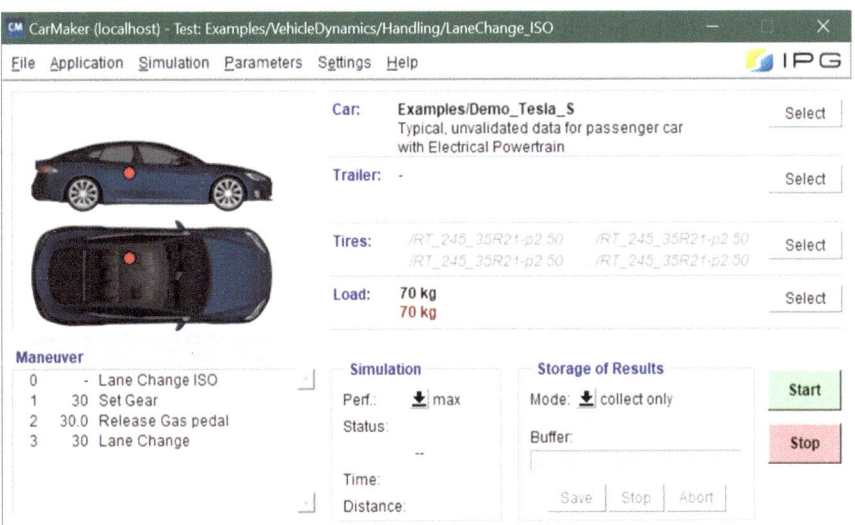

[IPG Automotive]

Fig. 1.5 CarMaker

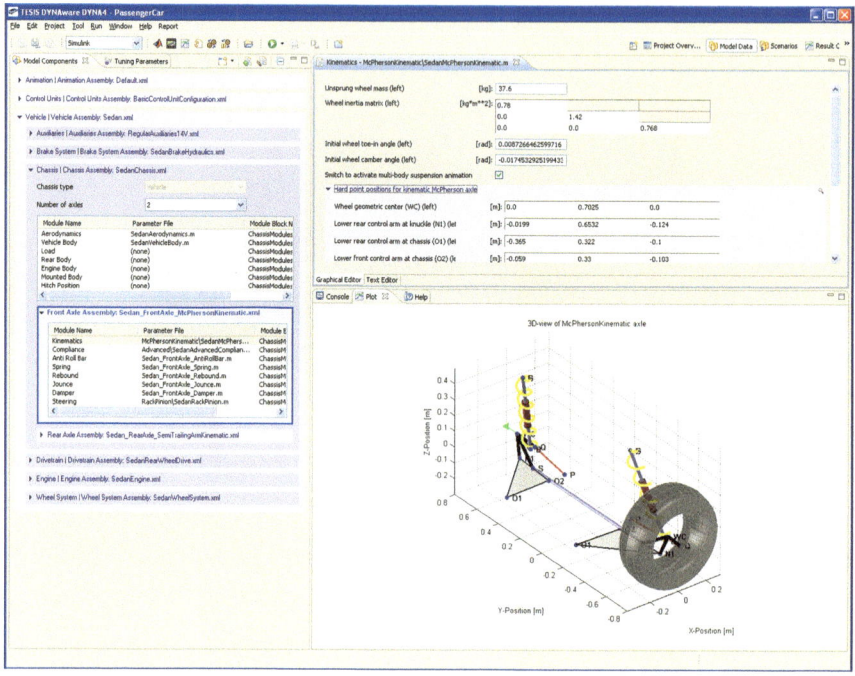

[Tests Dynaware]

Fig. 1.6 DYNA4

Further explanations on this topic can be found in general on multi-body systems for example in [AnZa98] and [Woer11] or more specifically on vehicle applications in [PoSc93], [RiSc10] or [ScHB10].

1.2.3 Block-Oriented Methods

The class of block-oriented methods has mainly arisen from system-technical approaches, as they are usual in control engineering. It is based on a signal flow diagram that can connect transmission blocks with a wide variety of functions.

In this environment, developers are therefore used to thinking block-oriented and modeling accordingly. Comprehensive analyses are often only possible with linear models that are integrated into a controller (Fig. 1.7). These systems can be modeled with MBS programs. Nonlinear models for a working point can be partially linearized. In [KoLu94] modeling techniques for linear models are described.

Here is certainly the most widespread program MATLAB/SIMULINK of The Mathworks. With this program package controllers of the most different complexity can be created. For the commercial MBS programs, there are usually interfaces to create controller models

Fig. 1.7 Simple control loop

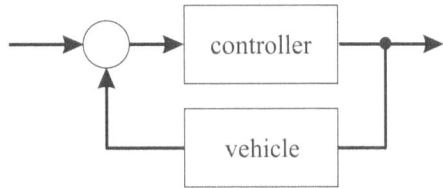

from MATLAB/SIMULINK or other comparable simulation environments. There are various ways in which this can be done.

The controller model can be exported as C code and connected to the MBS program via a suitable programming interface. Some MBS programs have the option of exporting the model as source code and then integrating it into the controller environment. Another possibility is co-simulation, in which either the MBS program or the controller environment is the master, i.e. it controls the numerical analysis and the program flow.

Further commercial program packages are for example MATRIXx by the company National Instruments or ASCET by the company ETAS. In the environment of the Open Source Community, the programs GNU OCTAVE or SCICOS can both cover sub-areas of the commercial tools.

Special applications in the field of hydraulic simulation are DSHPLUS by the company Fluidon or AMESIM of Siemens PLM Software. A large catalogue of hydraulic components is available in each case in order to be able to create a model that is as realistic as possible. The connection to an MBS program can often also be made using the methods described in this section.

The mentioned program packages are of course only a selection. Especially on the market of special applications there is a large number of tools, which partly deal with very special questions.

References

[Adam07]   ADAMSKI, D. ET AL: Rechnerische Lastkollektivermittlung auf der digitalen Straße, in
           *Lastannahmen und Betriebsfestigkeit, 34. Tagung des DVM-Arbeitskreises
           Betriebsfestigkeit*, 23-29, Wolfsburg, 2007
[AnZa98]   ANGELES, J. AND ZAKHARIEV, E.: *Computational Methods in Mechanical Systems*,
           Springer, New York, 1998
[Broc14]   Die Brockhaus Enzyklopädie Online, F. A. Brockhaus / wissenmedia in der
           inmediaONE GmbH, Gütersloh/Munich, accessed on 17.02.2014
[Klei12]   KLEIN, B.: *Grundlagen und Anwendungen der Finite-Element-Methode im Maschinen-
           und Fahrzeugbau*, Springer Vieweg, Wiesbaden, 2012
[KoLu94]   KORTÜM, W. AND LUGNER, P.: *Systemdynamik und Regelung von Fahrzeugen*, Springer,
           Berlin, 1994
[Neuw04]   NEUWIRTH, E. ET AL: Operativer Einsatz des Virtuellen Prüfstandes zur Darstellung von
           Betriebsfestigkeitsprüfungen an Gesamtfahrzeugen mithilfe der Mehrkörpersimulation,
           in *Berechnung und Simulation im Fahrzeugbau*, 381-408, Wuerzburg, 2004

[PoSc93]   POPP, K. AND SCHIEHLEN, W.: *Fahrzeugdynamik*, Teubner, Stuttgart, 1993

[RiSc10]   RILL, G. AND SCHAEFFER, TH.: *Grundlagen und Methodik der Mehrkörpersimulation*, Vieweg+Teubner, Wiesbaden, 2010

[ScHB10]   SCHRAMM, D., HILLER, M. AND BARDINI, R.: *Modellbildung und Simulation der Dynamik von Kraftfahrzeugen*, Springer, Berlin, 2010

[Shan75]   SHANNON, R.E.: Systems simulation: the art and science. Pretince-Hall, Englewood Cliffs, 1975

[Woer11]   WOERNLE, CH.: *Mehrkörpersysteme – Eine Einführung in die Kinematik und Dynamik von Systemen starrer Körper*, Springer, Berlin, 2011

2

Make things as simple as possible – but not simpler.

Albert Einstein

Systems engineering is a field of science with a large number of publications in a wide variety of fields. When studying engineering, some of the following terms are often used in control engineering. I will try to introduce them with regard to our main topic and to find illustrative examples of how to apply them in the chassis. As a minimum system, the simple linear spring runs through this chapter, which every engineer has been familiar with since the first semester of technical mechanics at the latest.

2.1 Concept of Systems

Since the term "system" is frequently used in everyday language, a closer look is required in order to be able to use it as a technical term. Colloquially, a system is often equated with political forms of society, but also with objects that are distinguished from their surroundings by their outer shell. However, systems in our sense are more abstract in nature, because real objects contain an immense number of properties that cannot and must not all be the subject of contemplation at the same time. A system, as defined below, is an abstract subset of the whole. Finding this subset is an essential task of systems engineering and modeling discussed in Chap. 3. The concept of systems originally comes from communications engineering, where attempts were made to separate it from the specific content of a message and to formulate this mathematically.

© Springer Fachmedien Wiesbaden GmbH, part of Springer Nature 2021
D. Adamski, *Simulation in Chassis Technology*,
https://doi.org/10.1007/978-3-658-30678-6_2

The Association of German Engineers (VDI) describes the term simulation in [VDI3633] as follows:

> Method for simulating a system with its dynamic processes in an experimental model in order to gain knowledge that can be transferred to reality.

This definition shows the central role the system plays in simulation engineering. The system is a self-contained unit consisting of one or more structurally connected elements. The state of the system may depend on other systems (or itself) and may affect other systems (or itself). To clarify what is inside and what is outside the system, a boundary—the system boundary—must be drawn, which is discussed in the next section.

The systems engineering makes use of the fact that also physically completely different systems follow the same laws and can consist of the same types of system elements. This leads to the possibility of systematizing similarities of these systems and describing them with equivalent mathematical relations. The aim is therefore to formulate general relations that reflect the system behavior as a function of its components and their structural connections [Boss89]. This does not mean that these systems are the same in terms of their elements or structure. But, it means that despite the structure and elements of each system, the same mathematical formalism can take place.

Perhaps, some of the readers have already encountered the analogy between the single-mass oscillator from technical mechanics with the properties of mass m, spring stiffness c and the damper constant d and the LC circuit of electrical engineering with the properties of ohmic resistor R, the inductance L and the capacity C (Fig. 2.1).

Both systems, as different as they are, can be described with the same system equations. The differential equations of the second order, which describe the system behavior of a free oscillator, are basically identical.

$$\text{Single} - \text{mass oscillator}: \quad f(x) = m\ddot{x} + d\dot{x} + cx \tag{2.1}$$

$$\text{LC circuit}: \quad f(i) = LC\frac{d2i}{dt^2} + RC\frac{di}{dt} + i \tag{2.2}$$

2.1.1   System Boundary

A system definition is an attempt to describe a real situation. The question with which the system is examined always plays a decisive role. Aspects that are not related to this question are omitted. If, for example, the driving behavior of a vehicle is described, the characteristic of the vehicle color plays no role in this context. It can be omitted in the formulation of the system. Other properties may well have an influence on the question, but their impact is rather small. Thus, these properties are also neglected in the system description in order to reduce the complexity and scope of the description. Whether a sun visor is extended or not,

Fig. 2.1 Mechanical single-mass oscillator (left) and electrical LC circuit (right)

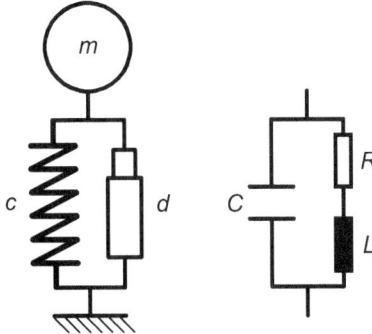

although the center of gravity of the vehicle will shift, there will be no demonstrable effect on driving behavior. Whether a retractable hardtop of a roadster is open or not, can then no longer be neglected with regard to the vehicle's center of gravity.

This process is also known as *separation* [Schw90]. The demarcation process is essential, because a system is determined primarily by what it contains and what it does not contain, i.e. what is outside the system boundary [Cell91]. The system environment is located outside the system boundary, but it must also be limited, since only the part of the environment relevant to the respective problem can be considered [Ropo75]. A strong interaction of the inner elements takes place within the system boundary. These elements interact only very weakly with the outside world. Nevertheless, information, energy or substances can enter and leave the system via the system boundary (Fig. 2.2). Environmental factors that are not directly influenced by the elements of the system are regarded as external system variables [Boss89].

If, for example, we take the wheel brake as a system, the heat generated during the conversion of kinetic energy is released into the environment and is therefore no longer part of the system. If the stability of the brake is to be investigated, an energy balance must be drawn up. If less heat leaves the system than is generated, the brake heats up.

If an entire vehicle is described as a system, the limit to the environment must be defined. While the tire will be part of it, the road is already outside the system. While the body shape with its aerodynamics belongs to the system, the air with its speed and direction will be outside the limit. The advantage in this demarcation lies in the fact that the same system *vehicle* can be used with different road or cross wind excitations. In many simulation environments, this is taken into account by the fact that the road information ($z(x, y)$, $\mu(x, y)$) and the wind (speed, direction) can be indicated independently of the vehicle model. But, sometimes not, because if road information is an integral part of the tire model, this may lead to an increased effort. We approach this problem in Sect. 12.5.

In addition to the border definition of what is inside and what is outside, the communication between the inside and the outside must also be described. Interfaces have to be defined, but their number should be as small as possible. Here, too, a selection must be made as to which interfaces are absolutely necessary for the respective question, which are unnecessary and which can be neglected. The different systems communicate with each other via these interfaces.

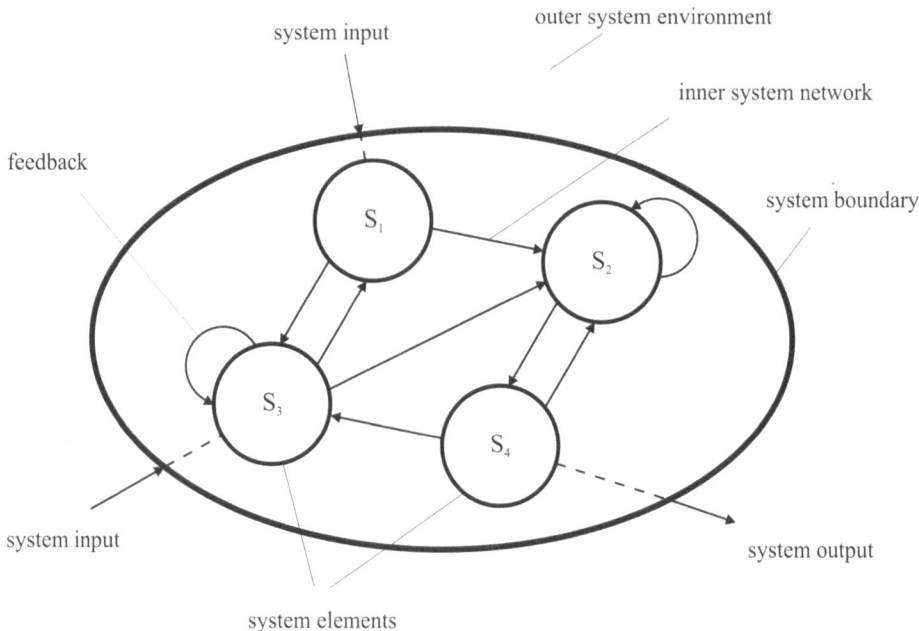

Fig. 2.2 System structure

2.1.2 Causality

The number of described interfaces will always be less than or equal to the real interfaces. One must make sure that even after a reduction, all interfaces are still *causally* connected. Causality means that the connection between cause and effect is not overridden. Nature itself provides for this requirement, but if one is to describe it with one's own words and algorithms, this can get mixed up. Here, above all, the temporal aspect is to be seen. An event C (cause) can, under certain conditions, be an event E (effect). This means that the cause C always needs to precede the effect E in time. E can never occur if C has not occurred first. A simple way to illustrate this is a simple arithmetic task, which is described in two steps (Fig. 2.3).

First of all, A must be available, from which the result B is calculated, which is required for the next step, the calculation of C. A calculation of C without B will not lead to the correct result. This connection is imperative. But, it is certainly not immediately apparent to everyone why this order is not always complied with *by itself*.

Those readers, who are able to create the system with their own program or program part using a higher programming language, know the so-called if-then-statement. With its help, you can create an if-then relationship. IF there is a particular situation (e.g. A > 0), THEN execute the following instruction (B = 2 A). The professionals know, of course, that the above example is incomplete. But, here lies a possible error that can contribute to the

Fig. 2.3 Causality of a math problem

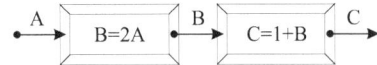

violation of causality. The complete relationship is called IF THEN ELSE. For our example, this could mean IF A > 0 THEN B = 2 A ELSE B = 0. The first approach contains an instruction only for the case that A > 0. If A ≤ 0, B would not be calculated. B would either not be defined[1] or an old value for B might be obtained from the last calculation step, which would be used instead of the correct value. The calculation continues to run and this may not even be noticed. If necessary, the argument could emerge that, in this special case, A cannot be physically less than or equal to zero because, for example, a mass is involved. This may be correct, but an error in the previous program flow could have caused a negative mass or the later user may have made incorrect specifications for the parameters used by the program. The approach applies here:

If something can be done wrong, it will be done wrong!

Those readers without programming knowledge must trust that the commissioned professionals take this approach to heart and test for possible errors.

Another way to violate causality is to parallelize the original serial process of the task. For reasons of saving computing time and the availability of corresponding hardware, complex calculations are split up, simultaneously calculated on different processors and reassembled afterwards. Of course, this is not possible for every type of task, but some are just imposing themselves for it. In these cases, it is important to pay close attention to whether there are dependencies between the parallel subprograms that could lead to a violation of causality.

2.1.3 Transmission Behavior

Systems engineering, as described, originally developed from a question of electrical engineering, especially communications engineering, in which the effects of the transmission of messages by the variation of current and voltage via electrical transmission devices were investigated. In classical systems engineering, the relationship between the input variables and the output variables of a linear, time-invariant system in the frequency domain was sought. This gave rise to the concept of the transfer function, which was also used in control engineering at an early stage. The focus is still on the reaction of a given system to external influences. These external influences are referred to as input variables. The system response to the influence of the input variable (cause) is obviously called the output variable (effect). Systems engineering tries to formulate the connection

---

[1]Which, depending on the programming language, could lead to a program abort and thus would amount in no *wrong* results or in a less convenient case, could amount in a random result.

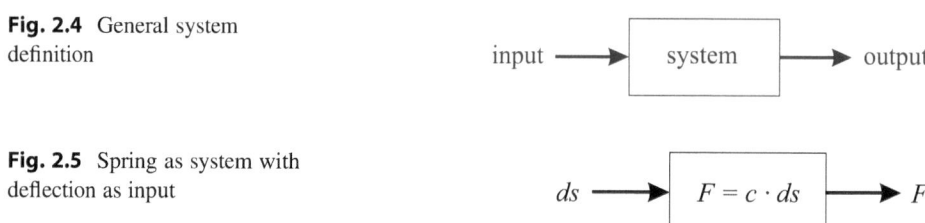

Fig. 2.4 General system definition

Fig. 2.5 Spring as system with deflection as input

between the output variable and the input variable of a dynamic system as generally as possible (Fig. 2.4).

As a concrete example, one can imagine a linear steel spring (Fig. 2.5). It has the simple force law $F = c \cdot ds$ if F is the force, c the stiffness and ds the deflection. The spring system therefore has a transmission behavior that is represented by its stiffness. If the input information consists of the deflection, it converts it into a force, which we then regard as the output variable.

Even for this simple example, a few agreements need to be made, in order to make it work. If we look at the real spring, we can apply a deflection and measure a force, but just as well we can apply a force and measure a deflection of the spring. The transmission behavior shown in Fig. 2.5 can only represent the first case. For the second we need a *new* system or a new modeling (Fig. 2.6).

Otherwise, we will not get the desired result. Thus, while the real spring has all its properties at the same time, the representation as a system with a transmission behavior is limited to the properties interesting to the respective question. Of course, it is software-technically feasible for the system to recognize the type of input information and then apply the correct form of force law. However, this does not fundamentally change the fact that you have to make the agreement what has to happen at the respective input information.

2.1.4   Range of Values

Another agreement concerns the value range of the input and output information. The question of the units is immediately visible. Is the stiffness c of our spring given in N/m, the deflection must also be given in meters. If it is incorrectly entered in millimeters, the forces are too great by a factor of 1000. The sign also belongs to the value range. Are deflections below zero allowed? In mechanics, it is generally agreed upon that negative deflections from the unstretched length lead to negative forces (compressive forces) and positive ones to positive forces (tensile forces). But, what if the input information is a time or mass? A mathematically formulated transfer behavior always provides a result, whether this then still corresponds to the real system, must also flow into the consideration of how to represent the system. Let us stick with our spring for now. The force law used so far only applies within a certain value range. If you squeeze the spring too much, it can run onto the block, i.e. the individual coils are in contact with each other. If they are pulled

Fig. 2.6 Spring as system with
force as input

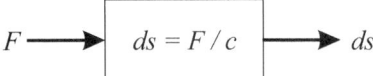

$$F \longrightarrow \boxed{ds = F/c} \longrightarrow ds$$

apart too far, the elastic area is left and plastic deformations occur. Both cases do not reflect the simple law of force. As a consequence, one must either limit the value range or extend the force law accordingly, otherwise one receives a wrong system behavior. The law of force also provides forces for spring lengths that are shorter than the block length according to the linear behavior. What linear and what non-linear behavior is, we will clarify in the next section.

2.1.5   Linear and Non-linear Systems

Systems are described as linear, if their system equation $y(t) = g(u(t))$ for any $u(t)$ and for all t has the following properties:

$$\text{Homogeneity}: \quad g(c \cdot u(t)) = c \cdot g(u(t)) \tag{2.3}$$

$$\text{Superposition}: \quad g(u_1(t) + u_2(t)) = g(u_1(t)) + g(u_2(t)) \tag{2.4}$$

If at least one of these two principles is violated, the system is non-linear.

Homogeneity means that a doubling of the input variable $u(t)$ results in a doubling of the output variable $g(u(t))$. If we look again at the spring from the last section, this applies to the formulated law of force. The working area of the spring, where it deforms exclusively elastically, is therefore linear. As soon as the windings go to block or the yield point is exceeded, this relationship no longer applies and the behavior becomes non-linear. At these points, the transition to another law of force takes place.

Superposition says that the effect (output variable) of a composite input variable $u_1(t) + u_2(t)$ must be the same as the sum of the output quantities for the individual input quantities. This principle makes it possible to represent larger systems by combining sub-systems. Section 3.9 uses this principle for the concept of modularization.

Engineers become familiar with linearization relatively early on. In general, the trigonometric functions are to *blame*. At the latest in the first semester of mechanics one hears that for small angles around a working point the following relationship applies:

$$\sin \alpha \approx \alpha \tag{2.5}$$

$$\cos \alpha \approx 1 \tag{2.6}$$

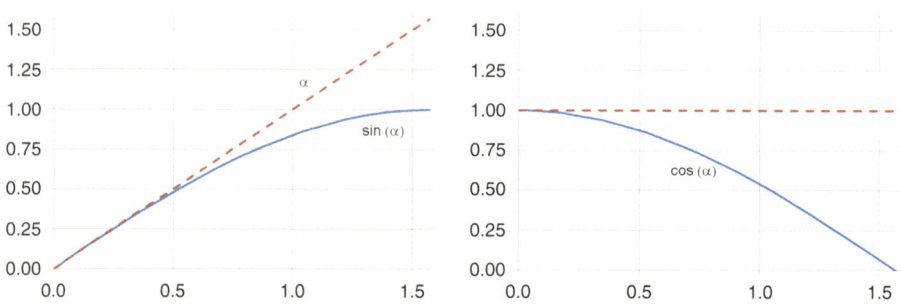

Fig. 2.7 Linearized and accurate course of trigonometric functions

One question is almost never answered: What are small angles and when is it no longer allowed to linearize? If you take a look at the course of the two functions from 0 to $\pi/2$, a strongly increasing deviation can be seen in the cosine (Fig. 2.7).

Of course, there is no general answer up to which range you can linearize—the accepted error depends as usual on the question. Unfortunately, we engineers often ignore such mathematical relationships and often linearized models are used far beyond their scope.[2] If an error of 5% is accepted, the sine could be used up to 0.54 (31°) and the cosine up to about 0.31 (18°). Since both functions are usually used simultaneously, the limit for using the linearized model would be just under 0.31 (18°) (Fig. 2.8).

If the system cannot reach larger angles due to its design, you are on the safe side. The problem of linearization does not lie in answering the question of what small angles are, but in the potential use of linearization for any angle. Once you have described your system linearly, you will want to use it for various applications. In doing so, one often loses sight of the basic idea, since this only applies to the direct surroundings of the working point around which one has linearized. In particular, if the model is used by third parties who may not be aware of this restriction.

Linearization makes sense in order to greatly simplify formula contexts. However, this is not always necessary with the current state of computer technology. An expert or a suitable computer algebraic program can often help your own knowledge of mathematics to take you a step further. Once this hurdle has been successfully cleared, the non-linear model approaches can also be used for *large* angles.

---

[2]All mathematicians and physicists, who would never do something line that, may feel vindicated now.

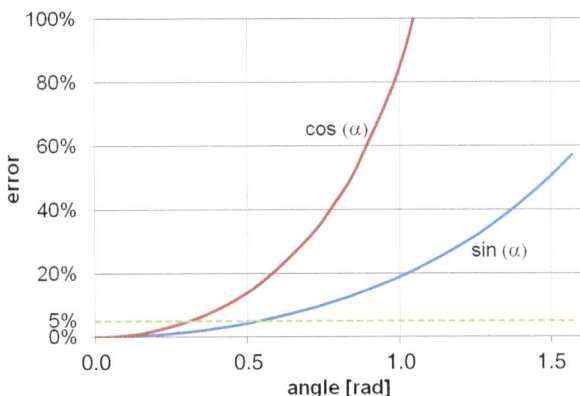

Fig. 2.8 Relative error of linearization

2.2   System Behavior

2.2.1   Systems with and Without Memory

When determining the time behavior of a system, it is important to distinguish whether the system has a *remembrance* or not. More technically, whether it has a memory or a direct coupling of the output signal to the input signal. If the time for the response signal is irrelevant, i.e. if the input signal is the same, the same output signal is produced again and again, regardless of when it is present and depending on what is happening, this is referred to as a static system. Our example from the last section, the steel spring, is such a static system. The formulated law of force has only one input variable of the deflection and has no internal memory that could take previous deflections into account.

Dynamic systems react differently, depending on the history of the input signals, because they *remember* their own history. In this way, a time behavior is implemented that corresponds to the current system state. Let us take the model of a damper as an example. The damping results from the fact that the kinetic energy is converted into heat. Since it is primarily a matter of viscous damping, it is mainly proportional to speed. The simplest consideration is a static system with a force law $F = d(v) \cdot v$. The prerequisite for this modeling is that the resulting heat is completely dissipated within each movement cycle and thus the viscosity of the oil in the damper remains constant. If the is not sufficiently cooled, the oil will heat up and the viscosity will drop. Thus, a damper with cold oil at the beginning of the analysis gives different results than one with heated oil at a later point in time. The system must therefore contain an energy balance to determine whether the heat can be released into the environment or whether it contributes to the heating of the oil. At least the result of this balancing must be stored and its influence on the viscosity implemented, then we have modeled a dynamic system.

Real systems are ultimately always dynamic, since they are always subject to change over a sufficiently long observation interval (e.g. ageing processes). Nevertheless, a static behavior of the system can often be assumed for shorter observation intervals if these changes can be neglected, i.e. if they have no direct influence on the system behavior of interest.

2.2.2 Change Behavior

For the input and output signals of the system, a distinction must be made between whether they change continuously over time (analog) or only at discrete points in time (digital). The difference is elementary. With time-continuous systems we know the system behavior at any time and for any further time with any temporal distance as well. The situation is different with time-discrete signals. They only change at certain points in time and between two adjacent points in time, we do not know what change the signal will undergo. Does it remain constant until the next time and then jump to the new value, or does the value change continuously? The greater the distance between two points in time, the greater the uncertainty about the system behavior between these points.

If, for example, we describe the system behavior with ordinary differential equations, we assume that the associated signal is a continuous process.[3] If we measure the system behavior with a digital measuring instrument, we only obtain information about the system state in the time grid of the sampling rate. This means that an originally time-continuous system becomes a time-discrete system due to the measuring method. If we look at an oscillatory system, the relationship between the period of oscillation and the sampling rate is decisive for what we get from the system behavior. In the processing of digital signals, the terms sampling theorem and alias effect are used for this purpose, the latter being evident in Fig. 2.9. If the system is scanned too slowly, an oscillation behavior is *measured*, which in no way corresponds to the real behavior. The sampling rate must therefore be a multiple of the natural frequency so that the true course of the signal becomes clear.

If we simulate the system behavior, we usually use numerical approximation methods as described in Chap. 4. Here, instead of the sampling rate of the measurement method, the step size of the integration method leads to the discretization of the time behavior of our system. Here, too, the step size must be much smaller than the time behavior we want to observe.

But not only the time behavior is discreetly represented. The amplitude of the signal is also discretized due to finite accuracy. For this purpose, the range of numbers to be considered must first be known, then it is converted into n equal parts when n is the accuracy to be achieved. Let us take as an example an acceleration signal whose value range we estimate to be ± 5 g ($\pm 5 \cdot 9.81$ m/s^2 = ± 49.033 m/s^2). We assume 8 bits as the accuracy, which means that we can display $2^8 = 256$ different values. If we divide the

---

[3]We only consider the technical mechanics here and not the quantum mechanics in which ultimately everything does change discretely.

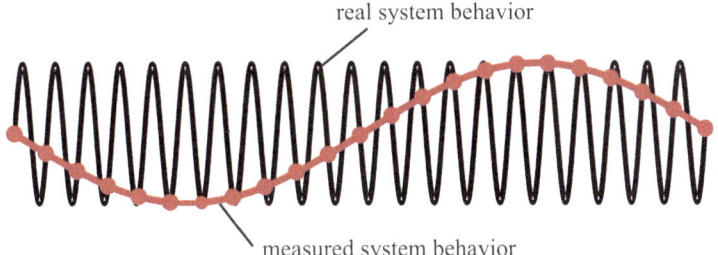

real system behavior

measured system behavior

Fig. 2.9 Sampling of an analog signal with too low sampling rate

signal's bandwidth of 10 g by 256 we get a resolution of 0.383 m/s^2. So we can only evaluate integer multiples of this value. If we are interested in smaller values, we have to limit the value range or increase the accuracy. A measured value that would be evaluated as number 17 then represents the value -49.033 m/s^2 + 17 $\cdot$ 0.383 m/s^2 = -42.521 m/s^2. How values less than -5 g or greater than $+5$ g are displayed must be defined.[4]

Since the simulations of systems take place on digital computers (even if everything actually began analogously), all calculated signals are also displayed discretely. With today's computer technology, an accuracy of 64 or even 128 bits is no longer a real problem. At the latest, when one deals with the integration of real digital control systems, one will also have to deal with lower accuracies. This means that the calculated signals are both discrete in time and value, which is illustrated in Fig. 2.10.

2.3   Questions from the Given System Structure

Depending on the given system components input signal, transmission behavior and output signal, different questions arise which are described below. In this book, I am almost exclusively concerned with the analysis of suspension components—nevertheless, I will briefly introduce you to the other two approaches.

2.3.1   System Analysis

The most common case for a simulation is the analysis. We know the system behavior because we have formulated the transfer behavior ourselves or know how it was formulated. So we know what the force law of our spring looks like. Then we know the input signals inducing our system. In the case of the spring, these could be different displacements. The output signals that the spring produces, when it is subjected to the

---

[4]For instance, one can generate an error (Overflow/Underflow) or use the maximal or minimal displayable value.

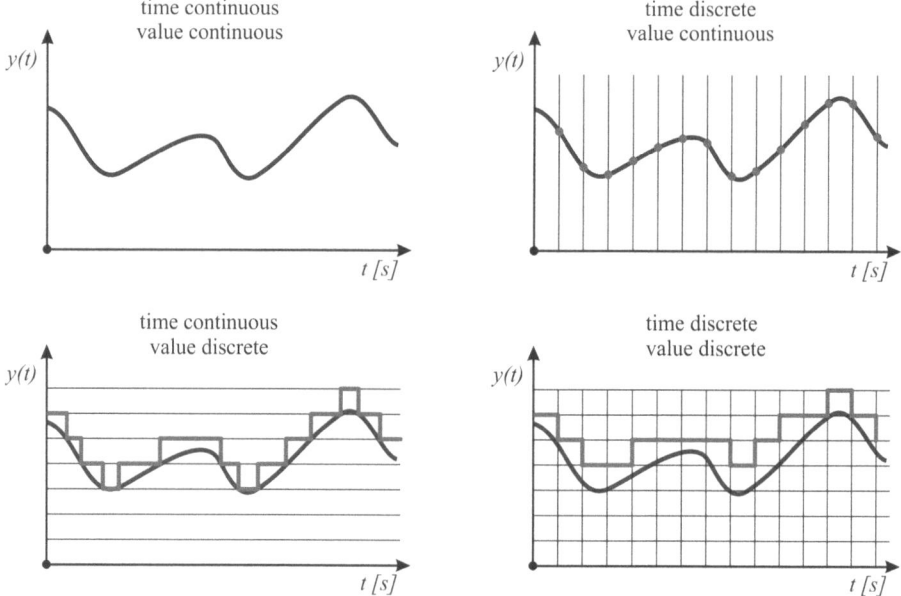

Fig. 2.10 Time and value discretization of a signal

excitations, are searched for. In the case of the spring (Fig. 2.11), the calculation can almost be saved, but it is hopefully understandable that in more complex systems the answer is not always obvious.

With this constellation, we expect a clear, reproducible answer. We know the system completely,[5] otherwise we would not have been able to describe it and therefore have a so-called white-box system.

2.3.2   System Identification

In contrast to the previous case, we do not know our system or do not know it completely—we have a so-called black box system. For example, we have a competitor's system on the table and want to know how it works in terms of benchmarking. Or we simply have no idea (or not the right mathematical description form) how to describe the behavior. We will try to identify the system behavior. To do this, we test the behavior of the system on different input signals by measuring the corresponding output signals. With the help of the procedures of system identification, we try to formulate a substitute system which shows the same behavior. So, we are looking for the so-called transmission behavior.

---

[5]According to the necessary completeness for this issue.

Fig. 2.11 Spring analysis

Fig. 2.12 Spring identification

Fig. 2.13 Spring control

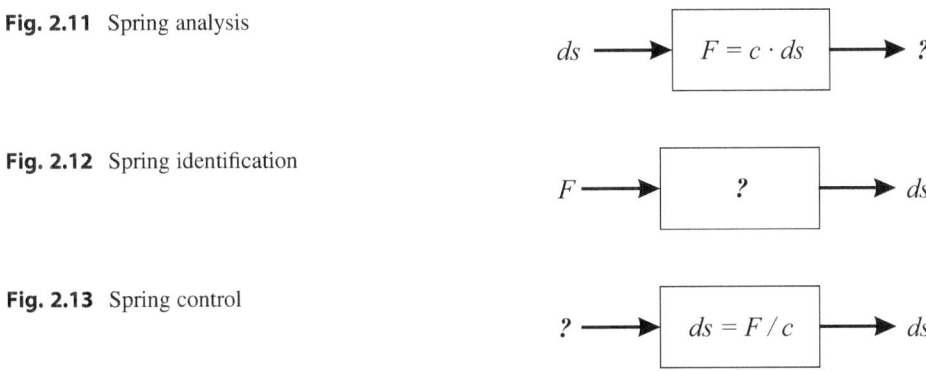

   With some experience and the right methods we will hopefully find a suitable substitute system. However, this system is directly dependent on the input signals we have used for identification. If we use the determined transmission behavior for other signal types or value ranges later, the validity of the substitute system can be quickly abandoned under certain circumstances.

   The substitute system is not unique, which means that it is *only* one possible solution. There can also be as many other solutions as you like that produce the same output signals for the selected input signals. Physical parameters of the real system can only be estimated if the basic idea of the substitute system takes them into account.

   If we were to identify our spring, we would be finished quickly (Fig. 2.12). A few deflections of the spring lead us to a straight line and the parameters of the straight line equations can be found with two points. Let us give it a try with an ABS controller.

2.3.3   System Control

In the latter case, which the combinatorics provides for three elements, we know the system behavior (white-box) and the desired output signal. So, we are looking for the right input signal. Depending on the system, there is exactly one or more solutions for the problem you are looking for. It is therefore a matter of correctly controlling our system, provided that this can also be achieved with the system. In the case of the spring, the question is what force we need to obtain the desired deflection (Fig. 2.13).

   If a certain potential energy is to be stored in the spring, the following applies

$$E_P = \frac{1}{2}\, c\, ds^2 \tag{2.7}$$

with which two possible solutions exist, since the spring can store the energy both as compressive force and as tensile force.

References

[Boss89]   BOSSEL, H.: *Simulation dynamischer Systeme*. Vieweg, Wuerzburg, 1989

[Cell91]   CELLIER, F. E.: *Continuous System Modeling*. Springer, Berlin, 1991

[Ropo75]   ROPOHL, G.: *Systemtechnik - Grundlagen und Anwendung*. Carl Hanser Verlag, Munich, 1975

[Schw90]   SCHWARZE, G.: *Digitale Simulation*. Akademie-Verlag, Berlin, 1990

[VDI3633]   VDI-Richtlinie 3633 (Draft): *Simulation von Logistik-, Materialfluß- und Produktionssystemen – Begriffe*. Verein Deutscher Ingenieure, Duesseldorf, 2013

Modeling

3

> *It is the mark of an educated mind to rest satisfied with the degree*
> *of precision which the nature of the subject admits and not to seek*
> *exactness where only an approximation is possible.*
>
> *Aristotle, Nicomachan Ethics*

3.1   At the Beginning, There Is the Problem

Modeling is certainly the most complex and important task when it comes to simulating technical systems in general and vehicles in particular. All mistakes made in this phase are passed through all subsequent steps and thus cause considerable effort and corresponding costs. For this reason, one should take enough time for the modeling and proceed carefully. So-called quick-and-dirty solutions often lead to fast results, but their quality is often not sufficient to make decisions for the product.[1]

Since the modeling depends very much on the task and the question, and on the real system to be modeled, there are few formalized procedures for this important step. Which is the appropriate physical or mathematical substitute model for the combination of question and system cannot always be answered unambiguously. Certainly the technical background of the modeler also plays a role. Anyone who works a lot with block-oriented tools, for example, will initially use this tool for chassis questions as well. Those who prefer to follow the ideas of technical mechanics will rather choose the method of multi-body systems or the finite element analysis (see Sect. 1.2).

---

[1]In the most inconvenient case, one cannot even estimate the quality of the results.

© Springer Fachmedien Wiesbaden GmbH, part of Springer Nature 2021
D. Adamski, *Simulation in Chassis Technology*,
https://doi.org/10.1007/978-3-658-30678-6_3

A criterion for the choice of the modeling method is of course the system itself. If the motion behavior and the forces and moments occurring in the chassis are to be recorded, the multi-body system method is certainly the best solution. If the component stresses or the acoustic transmission behavior are of primary interest, the finite element analysis is more likely to be used. If, for example, the transmission behavior is to be investigated for an active chassis control system, the block-oriented procedure mentioned above can be used here. Depending on whether the focus is on hydraulics, electrics or control, an appropriate tool can be used.

In summary, however, it can be said that the model must represent the essential system properties of interest and enable a statement to be made about the system behavior so that the respective question can be dealt with. Here, too, [VDI3633] contains a suitable definition of the term model:

> Simplified reproduction of a planned or existing system with its processes in a different conceptual or physical system. In terms of the investigated properties, the model differs from the original only within a tolerance framework that depends on the examination target.

In [Ammo97] the model is very appropriately called condensed knowledge, which is to be understood as the reduction of reality. Whereby reduction is meant here rather in the sense of cooking, that is, one separates the useful from the useless. Accordingly, the description of the essential properties within the meaning of the VDI regulation within the framework of a defined investigation objective is the same. The tolerance frame describes the accuracy of the statement required to answer the question.

The VDI regulation makes another distinction. Does the system to be modeled already exist or is it still a concept? The resulting derivatives for the modeling are very different. If the real system already exists, the purpose of modeling is usually to capture a previously ununderstood system behavior or to achieve an optimization of system parameters. The real system with its individual relevant characteristics must be mapped. It is often not sufficient to use nominal values of the target system. Instead, where necessary, measured actual values must be included in the modeling. At least as long as the tolerances of these values are considered essential for the quality of the calculation results.

If a future system is to be modeled, only a few system parameters are usually defined at the time of modeling and tend to reflect target values. Accordingly, the modeling may also be carried out roughly. If the used parameters are loaded with great uncertainty, it makes no sense to want to create a detailed model of high quality. The answers that can be expected from the calculations are rather indications of a direction and reflect qualitative behavior. A quantitatively correct behavior is to be expected in very few cases—this is reserved for later models with parameters of a higher level of confidence.[2]

---

[2]With the notion *confidence level*, I associate that the model or the underlying modeling technique have already been applied frequently and have been compared to the real system afterwards.

However, knowledge of the system to be modeled is essential. What are the important properties, how do they depend on internal and external subsystems? What information is transported out of the system and what information is transported in? What cause-effect relationships can be established? If this knowledge is lacking, no suitable model can be created or later validated.

3.2   The Difference Between Erroneous and False

A substantial and here not hair-splittingly meant difference exists between erroneous and false models or results. Whenever one wants to represent a real or future system by a model, one neglects properties which one considers insignificant for the question. So, there are differences between the system and the model. The comparison of a luxury limousine worth a six-figure Euro amount with a two-mass oscillator does not seem admissible, but can be completely sufficient for some questions on the basic tuning of springs and dampers.[3] A modeling error is thus made when changing from the physics of the real system to its mathematical description. The danger, that relevant effects are disregarded or variables considered unimportant are neglected, can always be the cause of a deviation between the model and reality [KrNe98]. Moreover, in Chap. 4 we will see that the equations provided by mathematics cannot always be solved exactly. We need to numerically approximate them. These approximation methods also have their own errors, which we will divide into a relative and a global error. In addition, we use a digital computer that has a finite representation accuracy of numbers, which adds the so-called rounding error. The model needs parameters that may only be estimated and thus provide further deviations from the behavior of the real system. Some errors that can be made when comparing calculated results with measurements are listed in Sect. 3.8 and finally, the result can also be interpreted inaccurately. You see, a variety of ways to make mistakes. But all of this we summarize under the term *erroneous*.

The models that actually contain inadequate or incorrectly formulated system descriptions should rather be classified under false models. If the vertical suspension behavior of the vehicle is to be calculated with a single-track model or the transverse dynamics with a single-mass oscillator, then this is the wrong model class for the planned investigations. If, for example, the calculation of a spring characteristic curve from the actual kinematic point of action to the wheel center was miscalculated, for example because one forgot that the spring transmission is considered square and not linear, then an error was made which actually leads to incorrect results, since the spring is most likely too soft.[4]

---

[3]Don't worry, the fine tuning will be done using a real vehicle, which, in the beginning, however, might have already contained a tuning variant that has been enhanced by a distinctively more complex model.

[4]Those who do not know where the mistake is here, should read Sect. 9.4—everyone else may too, of course.

There are even supposed to be simulation programs which still contain errors in the software. You should not always assume this immediately if a result does not come out as expected, but you can keep it in mind.

It is not at all problematic to work with erroneous models in simulations—it is even almost impossible to work with errorless models. You only have to be aware of this fact at all times, i.e. the limited informative value of the modeling of the underlying questions and system descriptions must be known and taken into account. Statements beyond this boundary are marked by strong optimism[5] and are more prophetic than objectively justified. A good example are linearized models, which due to their linearization around one working point are only valid in the immediate vicinity of this working point. If the range of statements is extended far beyond the working point, extreme misjudgments can result. As long as the model shows extreme results that show everyone that something must have gone wrong, only manageable power dissipation has been produced. However, if the results are able to be interpreted as *right*—and almost everything can be explained—even larger financial damages can be caused. At the very least—and rightly so, in this case—the reputation or credibility of the computational engineers has been greatly diminished.

3.3   Methods for Modeling

3.3.1   Induction

The induction method is also known as the top-down method. One moves from the abstract to the concrete. This involves a step-by-step refinement without, however, changing the external appearance of the overall system (Fig. 3.1). With this procedure, you can only start modeling once the system has been fully described and understood.

In our question, for example, the overall system would be a chassis model for vehicle dynamics simulation. In the next step this is disassembled into its components (tire, suspension, spring, damper, steering, etc.). In the next level of detail, the tire subsystem is subdivided into the subsystems tire contact area, force transmission or mass properties, which in turn can contain the calculations of these sub questions with the focus on vehicle dynamics simulation. You get a hierarchically very strongly structured overall model for the underlying question. This approach is very obvious and is basically implemented in most programming languages. Here, a complete program is often formed by subroutines with corresponding sub calls, whereby the subroutines are to deal with the sub problems.

---

[5]Simulants are optimists by nature. Would they try explaining the world in calculation models otherwise?

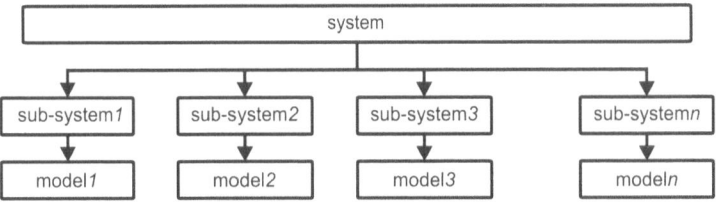

Fig. 3.1 Method of induction

3.3.2   Deduction

The method of deduction, also known as the bottom-up method, proceeds the other way round. Initially, unspecific basic components are generated, which also apply under general initial conditions. Together, these form a kind of construction kit which can be used to assemble a complete system (Fig. 3.2). You put the cart before the horse, so to speak. The advantage is, that concrete modeling ideas of partial problems can be approached directly and the energy of the creative moment can be used. The disadvantage of the method is, that concrete modeling ideas of partial problems can be approached directly. The energy of the creative moment sometimes makes you start full steam in the wrong direction. Without a complete overview, it becomes difficult to find a form that is as universally valid as possible. Experience has shown that many people find it difficult to turn away from their chosen path.

For our chassis example this means that a catalogue of component models exists. There are, for example, several suspension models which can then be used by specific parameters for the respective problem. Thus, rather basic principles are used which specify the general nature of the model describing systems of equations. This shall apply in the same way to any other chassis component.

3.3.3   Method of Choice

The method of choice, of course, does not exist. As so often, it depends on the overall context and personal experience on how you will proceed. Over time, however, it will certainly be a mixture of both methods. One creates a basic structure with induction and then deductively prepares sub model for sub model in such a way that they can be used not only for special but also for general questions.

When creating a model of a double wishbone suspension, the general transmission behavior is determined by the kinematics of two wishbones and the tie rod. Through a vehicle-specific parameterization of the kinematic points, the control arm lengths and masses, as well as the bearing stiffness, the general model becomes a concrete, vehicle-related model.

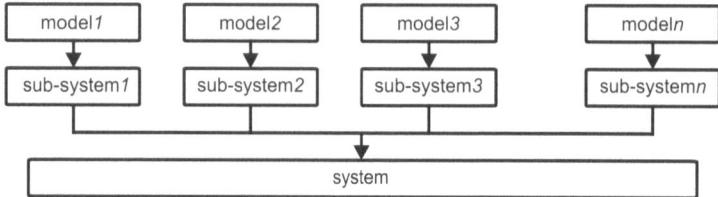

Fig. 3.2 Method of deduction

3.4   Model Classes

3.4.1   Physical Models

There are two main classes of models to be distinguished. On the one hand, physical models are used which contain a description of the system behavior via substitute models. These descriptions are usually idealizations of physical laws. Thus, for example, the behavior of a spring, as already introduced, can be determined by the simple correlation $F = c \cdot ds$. There is a functional relationship between the system parameters, here, between the stiffness c and the system behavior. The system parameters have to be determined by measurements of a real system (or estimation if the system does not yet exist). We will continue to deal predominantly with this class of vehicle models.

3.4.2   Behavioral Models

On the other hand, empirical models or behavioral models can be used. In [Ise192] they are called experimental models. They do not represent the physical components of the system, but their behavior. There is therefore no functional relationship between the system parameter and the system behavior. In order to stay with the simple system of the spring, for example, there is a component whose parameters are not known (black box). However, this component was measured and the input and output signals are present, so that the system behavior can be deduced by means of identification. For the spring, this will lead to the previous characteristic curve parameterizing a linear replacement model, but the experimental model has no stiffness parameter.

   For more complex components, this process is more elaborate. The replacement model must then be parameterized in such a way that it reflects the transmission behavior of the real system. It must therefore supply the same output signals when exposed to the input signals from the measurement. Although, *same* is a very elastic term. Let us leave it as it is that the correspondence between the output signals of the real and the modeled system must be sufficient for the required quality of statement of the task. In contrast to the physical models, which can also affect future systems due to the laws used, the model must be

Table 3.1 Selection of properties of physical models and of behavioral models

Physical models	Behavioral models
The model structure follows natural laws.	The model structure must be assumed.
The description of the behavior of internal state variables and the input/output behavior.	Only the input/output behavior is identified.
The model parameters are specified as a function of system parameters.	The model parameters are pure numerical values which generally do not show any correlation with the physical system parameters.
The model is valid for a whole class of a process type and for different operating states. However, many process variables are often only inaccurately known.	The model only applies to the process under investigation for a specific operating condition. But it can describe this behavior relatively precisely.
The model can also be created for a non-existent system.	The model can only be identified for an existing system.
The essential internal processes of the system must be known and mathematically describable.	Internal processes of the system do not have to be known.
	Since identification methods are independent of the individual system, a single identification software program can be used for many different systems.
Usually requires a lot of time.	Usually requires relatively little time.

available for experimental models. It must be possible to make measurements so that the input/output behavior can be approximated.

This model class can be used if the physical parameters play no role at all. One does not want to know what the actual component stiffness of the spring is, one only needs to know the behavior around one working point. Then it may be possible to fathom this external behavior of the system by measurements much faster than to analyze the individual components, i.e. the internal behavior. Another application could be to map the component or vehicle of a competitor. Normally, one does not know the inner behavior here and has to be content with the identification of it. In [Ise192] and [Ise292] different procedures are described which can be applied depending on the type of system. The comparison of the two model classes shown in Table 3.1 also originates from there.

Both model approaches can of course also be used mixed in a complex chassis model. Components whose behavior can be represented physically and for which corresponding measurements of the system parameters are available are used alongside components which are modeled as behavior models and whose substitute system parameters were found with the help of identification. This procedure is described in detail in [HaHo03].

3.5 Problem Analysis

At the beginning of the modeling, the problem analysis must stand. It is important that both the system to be considered and the question behind it must be analyzed.

3.5.1 Analysis of the Question

First of all, it must be clarified beyond doubt which question is to be dealt with at all and in what quality and type the answer is expected. How exact does the answer have to be? Is it a matter of a trend-setting decision or a concrete component design?[6] This point is often dealt with too briefly, because everyone involved knows exactly what it is all about. At this point I would like to refer to the relevant literature on the topic of *basics of communication.* Misunderstandings often occur here, which only become apparent much later, for example when the desired answer cannot be provided in the presentation of the results.[7]

3.5.2 Analysis of the System

Once the question has been clarified, the system itself must be discussed. How much do we know about the existing or planned system or its behavior? Which system properties are relevant for answering the question? Is it possible to generate at least one test case where an expected system response can be obtained with known input data? This serves to check whether the model created later actually represents the desired system. If the system already exists, measurements can be defined here, for example, with which the model can later be compared.

The answers to these questions form the basis for the later model design. Everything that was forgotten or inadequately answered in this phase is missing in the model or is not adequately mapped.

---

[6]In order for no misconceptions to arise: Both are important decisions and the trend-setting often contains a much greater financial risk. However, it is usually met rather early, when still relatively few data for the vehicle or the component are known. Therefore, all participants have to be understand that the statements of the simulation results are only qualitative.

[7]This does not concern the answer's content, but the kind of examination.

3.6   Model Design

3.6.1   Simulation Method

After having been clarified in the previous phase, *what* is to be modeled, the question now arises, *how* the system is to be modeled. The *How* includes the question of the method. If stresses in individual components are to be investigated, the finite element analysis will be used. If the non-linear motion behavior is more in the foreground, the method of multi-body systems should be used.

The focus of this book is on the latter, i.e. the system behavior in the time domain is of interest. In technical mechanics one learns that time behavior is represented by equations of motion, which usually consist of differential equations. In addition, there are also kinematic or physical binding equations. All together results in a larger system of equations, which can usually no longer be calculated by hand. Especially since in most cases, it will no longer be analytically solvable and therefore the help of numerical approximation methods will be needed. For the solution of this system of equations, a multitude of procedures have been created in the past, which are contained in different forms in the available simulation environments.[8] For example, [Woer11] provides a good deepening of this problem for those interested.

If a commercial simulation environment has already been purchased, the question of how to solve the equation no longer arises. Rather, it is the depth with which you want to map the system and which data you have available for parameterization. In the case that a simulation environment is still required, please refer to the explanations in Sect. 5.1.

3.6.2   Implementation of the Problem Analysis

In the previous step, the system was analyzed with a view to the problem, and system boundaries and system properties, which are essential for answering the problem, were worked out. This must now be implemented. The relevant properties must be represented by appropriate modeling approaches. The simplifications result in limits within which the model may be used. These limits and the necessary operating conditions should be described so that everyone who wants to use the model knows what is possible with it and what is not.[9] In this step it becomes concrete for the first time. After the previously theoretical preliminary considerations, a model must now be created. The success of this step depends significantly on the experience, expertise, system understanding and abstraction ability of the modeler. It limits the model quality. But experience can only be gained if

---

[8]The equation system's solution (Sect. 5.2.2) should not be confused with the numeric integration (integration process) in Chap. 4.

[9]An example of this would be the use of linearized models, as described in Sect. 2.1.5.

you do something and accept that the first steps in this environment do not necessarily always lead to the optimum. However, in order to be able to use the experience, the whole chain has to be maintained until the validation described in Sect. 3.8. This is the only way to achieve model and modeler improvements.

3.7   Verification

The first step in verification is to clarify whether the model is formally correct. This begins with the question of whether it corresponds to the question or specification at all. So, was the *right* model used? In addition, it must be noted that no sign or unit errors are contained and that all necessary components have been taken into account.

If, for example, a measured damper is to be simulated, it must be checked that the characteristic curve belonging to this damper is used and that the spring and/or tension stop possibly present in the real component was also modeled. Even if these things sound trivial, they are responsible for a large number of false calculations.

In order to really verify a model, there should be standards for naming and storing data or parameters. So that it becomes immediately obvious when a wrong record is used. The fact that the parameters are separated from the model is a procedure from the programming technique, which will be discussed again in Sect. 3.9.2.

A very helpful standard, which unfortunately is not offered by all simulation environments, is the use of units, which will be discussed later in Chap. 7.

3.8   Validation

The validation of the model at the end of the model formation represents the targeted comparison between the calculation results and measurements of the real system. Only a successful comparison can subsequently clarify the validity of the model for the question. This gives rise to a modeling principle that should always be observed for the productive use of simulation.

Each and every (!) model must be validated - models that cannot be verified do not need to be created.

3.8.1   Basic Procedure

Let us assume that the task is to create a full vehicle model of an existing vehicle in order to be able to carry out vehicle dynamics simulations. The model is to contain various component models that represent, for example, the steering system, the brake system, the drive train, the tires and the driver. Proud to have finally completed the vehicle model, one compares the results of the full vehicle simulation with the measurement results of the real prototype.

Combinatorics now allows for two cases:

1. The calculation results match the measurement.
2. The calculation results do not match the measurement.

The second case is clear—something is wrong with the model. And the first case—the match? Now here, too, we have two options again.

1. The model is really perfect and can reproduce the desired load case very well. We won!
2. The model is flawed and unfortunately the flaws overlap to such an extent that the desired result seems to be achieved. Lost!

What am I getting at? Starting with the full vehicle simulation without being sure of the quality of the component models is a gamble and not advisable. Deviations between the calculations and the measurements can in very few cases be directly assigned to one of the component models. If, for example, acceleration amplitudes coincide in the time domain, there may still be differences in the frequency domain. If the maneuver to be investigated lasts longer, an average effect is achieved, especially in the case of individual deviations that could become visible, for example, due to a threshold crossing, if integrating variables are considered over the entire period.

When considering which component models (and in which depth) to make, one should already have in mind which test scenarios are useful for the models. How can I check the function of the component model in the range relevant for the planned load cases? Can measurements from rig tests be used? For the component models described in the second part of this book, it makes sense to simulate the measurements of the components in the simulation. If, for example, a chassis bearing is measured statically and dynamically, the measuring procedure should be simulated if possible. This way it can be checked, whether the model behaves in the same way as the real component or at which points the model reaches its limits. In anticipation of Sect. 8.3, it should be mentioned that simple elastomer bearing models, for example, can reproduce static measurements very well, but often fail in dynamic measurements due to the frequency and amplitude dependence of the elastomer. This does not mean that such models should not be used, but that one must be aware of this and clarify the relevance of these effects for the respective problem or load case.

If measurements of component or full vehicle test benches are available, this can also be the basis of a validation. Depending on the complexity of the test bench or the measurement specification, it may be necessary to replicate the test bench, at least in part, as a model. Of course, this test bench model must also be validated, but if measurements are made there again and again, the effort can quickly pay off. In the case of full vehicle test benches (Fig. 3.3), the following applies within the framework of validation: Only when the individual models are certain, should one approach the full vehicle.

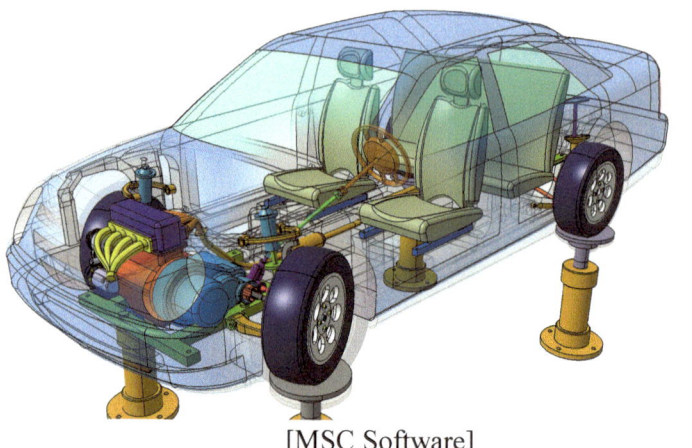

[MSC Software]

Fig. 3.3 Model of a four poster

At this point it must be added that the replay of the test or measurement signal used for modeling is not a validation. This result has already been used for modeling and is implicit in the model. Only a new signal can be used for validation.

3.8.2   Comparison of Measurement and Simulation

The first choice for a validation should always be a comparison of the calculated results with measured values on the real component. Classically, the real component is not yet available at the time of modeling, so that one first dares to make a prognosis for the component or vehicle behavior. As soon as the real behavior can be measured, the calculated variables have to be compared with the measured ones. This is the only way to learn about the quality of your model and the underlying modeling technique. Of course, you can also check the basic behavior of predecessor components in advance, where you then have to know the difference between the predecessor and the current component in order to be able to assign any deviations.

Let us first assume that measurements of the modeled component are available. Often, one gets measurements of components or vehicle measurements as a computational engineer without knowing exactly how these were created. Whether a zero adjustment of the sensors took place, whether the measurements were post-processed (filtered, scaled, kinematically converted, etc.), which boundary conditions prevailed and where, for example, sensors were installed. This makes the comparison difficult and adds some systematic errors to the validation.

As a first example, acceleration signals of the wheel in z-direction measured during a rough road crossing. The statement *wheel accelerations* can mean that the wheel has actually been measured, which is rather unlikely due to the rotation. It may also have been measured on the wheel carrier. The position of the real sensor depends on where it is placed. This can be dependent on the suspension concept in the most different places. If there is no space on the wheel carrier, a control arm is used instead. All of this can mean that a wheel acceleration has been recorded. The first mistake one can make as a computational engineer is to take the term too literally and use the acceleration at the wheel center. In the model, this location is usually available as a coordinate system and the accelerations at this point can be evaluated. Even if this point naturally also exists within the real wheel, there cannot be measured.

It is worth it, no, it is imperative to get an idea of how the sensors are attached to the vehicle with the colleagues from the measurement technology department. Afterwards, it should be decided how the signal is evaluated on the model. Those who have not yet understood why it does not matter where the signal is picked up, should remember their dynamics lecture on technical mechanics. Rotational accelerations are multiplied by a lever to translational accelerations. If one imagines the wheel suspension in the simplest case as a pendulum, an acceleration signal can be scaled with the different lever lengths of the discussed sensor positions. Thus, as can be seen in Fig. 3.4, the same acceleration $\ddot{\varphi}(t)$, depending on the position of the sensor, can lead to three different translational acceleration signals.

$$(1) \text{ on linkage} \quad a_1(t) = \ddot{\varphi}(t) \cdot l_1 \tag{3.1}$$

$$(2) \text{ on wheel carrier} \quad a_2(t) = \ddot{\varphi}(t) \cdot l_2 \tag{3.2}$$

$$(3) \text{ at the wheel center} \quad a_3(t) = \ddot{\varphi}(t) \cdot l_3 \tag{3.3}$$

In this simple example, the signal is scaled with the lever length between the instantaneous pole and the sensor position. On the real vehicle or on the model, the instantaneous pole position is usually not easy to determine, so that the scaling factor is not immediately available. In our example, all points of the wheel suspension move on a circular path relative to the instantaneous pole. This means that the sensor is deflected from its original vertical position and no longer only measures the z-component of the acceleration. If a coordinate system is evaluated in the model at the same position, which is also swiveled, the result is the same *error* and the signal remains comparable. Some simulation environments offer a type of coordinate system whose orientation always remains collinear to the inertial system. If such a system is evaluated, it is not rotated and there is a deviation from the measured signal.

A second example shows the sensitivity of the measurement and calculation signal to component and assembly tolerances. If the track curve of an axle is recorded,[10] its course is

---

[10]This concerns the course of the toe angle when the wheel compresses and decompresses.

Fig. 3.4 Measurement of wheel
acceleration with different
sensor positions

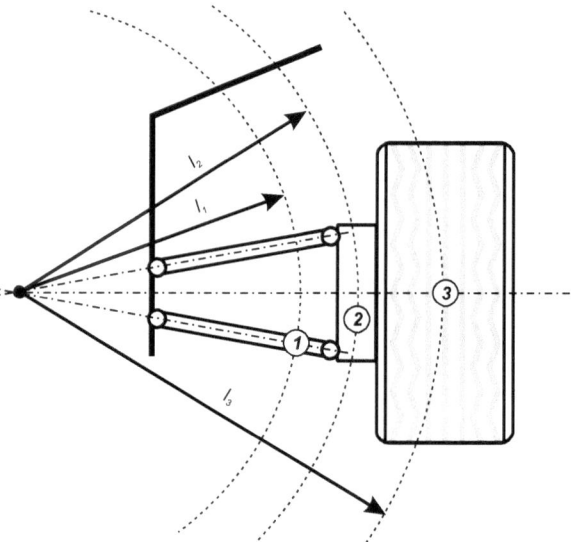

directly dependent on the position of the connection points of the tie rod. Even deviations
from the target position within the usual tolerances can be detected in the measurement.
The following applies again: The measurement is subject to tolerance, the calculation is
usually not. If such a significant dependency is known, it can be helpful to compare the
derivative of this curve around the respective zero position. In derivations, constant offsets
are eliminated and the change is valuated. If measurement and calculation are now the
same, this procedure indicates a deviation within the tolerance range. If they are not, there
are further differences to the real axis in the model.

A popular formulation when comparing two signals (e.g. a measured and a calculated
one) is that they are *line width* identical. In other words, they lie on top of each other. The
goal of every simulation to depict reality has therefore been achieved. A fallacy in several
respects. On the one hand, the requirement that measurement and calculation must lead to
identical results is fatal. It ignores the fact that measurement is only one of many possible
representations of reality. The repeatability of measurements depends strongly on the given
boundary conditions. If these can be reproduced exactly over and over again, the measurement results will be very similar—but hardly identical. The more difficult it is to record and
repeat the boundary conditions, the greater the scatter of the measurement results. If you
measure several times, you will get a set of curves of results and the demand on the
simulation results can only be that they lie within this set of curves and have the same
qualitative course.

Component tolerances lead to the same effect, which also lead to scattering. A nominal
component is usually calculated in the simulation. Tolerances or changes to the nominal
values due to wear or overloading are often not taken into account. Here, if the effort is
justifiable, at least the best and the worst component should be considered in terms of the
upper and lower tolerance limits. This way one can estimate the band in which the expected

results must lie. Influences from ageing or temperature can be investigated in a similar way. If it turns out that the influences on the final result are negligible—all the better. If not, one knows which scenarios one should examine in the future.

The only recommendation that can be made here is to coordinate closely with the measuring colleagues. How is measurement performed, where are the sensors, how is the raw data processed and how is it evaluated? The resulting answers must ultimately be used to map the measurement process in the simulation so that the results produced there run through the same process.

To sum it up: During validation, a solitary vehicle with component and assembly tolerances, with generally unknown external perturbations and an erroneous measurement, is compared with an erroneous nominal vehicle model and a (discretization) erroneous calculation result.[11] This should give the computational engineer sufficient support to tolerate the thesis that only measurements can show the true behavior.

3.8.3   Comparison of Simulation and Simulation

If no measurements of a real component are available, the behavior of the model can also be checked for plausibility on the basis of literature values. Validation can then no longer be considered in the true sense of the word. A common strategy is to measure the new model against older, perhaps more complex, hopefully validated models. For this procedure, however, it is imperative that sufficient experience with the reference models and a strong confidence in the result quality for the considered load cases are available. Otherwise, one only proves that the new model can reproduce the error of the reference model.

3.8.4   Validation with Full Vehicle Measurements

If the full vehicle behavior is to be simulated, the full vehicle model must ultimately be validated. However, this can only be the last step after a series of component validations, as explained in Sect. 3.8.1. If there are differences between the measurement and the calculation, it is usually not possible to assign these deviations to a specific component and, if necessary, two model errors may even compensate each other in the specific maneuver. The number of component and assembly tolerances in the real vehicle is so large that for this reason alone there can be a sufficient number of deviations from the nominal model. If the vehicle is more a prototype rather than a series-produced stand, all components of the vehicle relevant for the load case must have been measured. Otherwise, the cost of validation cannot be justified. Deviations between the measurements and the calculations

---

[11]The difference between erroneous and false has already been discussed in Sect. 3.2.

can then not only point to inadequacies in the model, but also to a lack of knowledge of the vehicle used.

The full vehicle validation is something like the supreme discipline of validation and you need some experience to be able to carry it out successfully.

3.9   Single or Multiple Use

3.9.1   Modularized or Monolithic?

If you are faced with the task of simulating something for the first time, you will make one model of the system to be investigated. The emphasis here is on the word *one*. One usually starts with a so-called monolithic model, which means that all necessary system properties are combined in one single model. This approach is obvious, because at this moment you only have the current problem in front of your eyes. If one compares modeling with programming in a higher programming language, a single program would result. If one really only has to deal with this question once and no similar one will come up to one, then nothing can be said against this type of modeling. But this is rarely the case. One will have to deal more frequently with the question and one will want to calculate variants again and again. For this reason, it makes sense to deal with the modularization of your model at an early stage. A module is a stand-alone, separate unit which can be assembled into a complete model via interfaces with other modules. So it is some kind of construction kit. The appropriate keyword is reuse. The rack-and-pinion steering model can thus be used for several vehicle models without having to be copied or rewritten each time. Depending on the engine, the Volkswagen Golf VII model is available with a twist beam axle or a multi-link axle as the rear axle concept. It is good to be able to replace the rear axle as a whole in the vehicle model. However, this type of modeling requires a relatively large initial effort, which is usually avoided, and so huge models are often created in the beginning that cannot be maintained over time.

What has to be done at the beginning of modularization? First, you have to be clear which kind of modularization the simulation tool supports. Here, especially the question of how many substructure levels are supported and how they can be connected is meant. If the root of the model is the full vehicle or body, the next layer or branch could be the front axle. Is it now possible to attach the wheel module to the front axle module or does the simulation environment only allow one layer, which must always be attached to the main module? Then the wheel would have to be integrated into the axle module. This is an unfavorable solution, since different wheels can be mounted independently of the axle, whose rims have different masses and mass moments of inertia. The same applies to the brake system, which usually grows with motorization.

The structure shown in Fig. 3.5 can of course only be exemplary, since it depends strongly on the basic vehicle structure. The dotted lines represent optional connections,

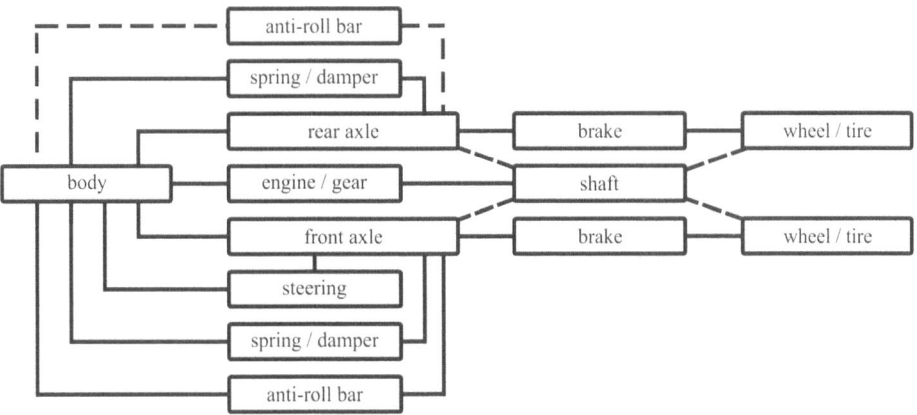

Fig. 3.5 Modularization of the chassis

e.g. for front, rear and all-wheel drive vehicles or vehicles that do not have an anti-roll bar on the rear axle.

The structure shown is still very abstract. Only the term *front axle* is used. The kinematics involved are irrelevant at first. The connecting lines represent the interfaces between the modules and involve more effort than one would initially assume. What belongs in a module and what does not? Should the top mount be part of the spring-damper module or the body module which also contains the body-in-white? A help for the decision can always be to consider from whom the necessary data for the parameterization come and whether there are forced links. The top mount will be based on the spring/damper design and less on the body shape, so that the decision will be relatively simple. Of course, the top mount could also be a module in its own right. Each real component could be represented by its own module. Whether this serves the overview and can still be administered meaningfully, must be critically questioned. However, it is often difficult, especially for newcomers, to get an overview of which module contains which component properties.

Once the decision has been made which properties should be contained in the module, the interfaces to the neighboring modules must be defined. One can orient oneself at the real connections or the force paths between the components. For the connection between the module *front axle* and the module *vehicle body* for example, all the connection points between the body and the control arms or between the body and the sub frame must be described. However, it should be noted that the abstract module *front axle* must represent multiple kinematics. A double wishbone axle has four joint or bearing positions per wheel suspension, a McPherson axle only three.[12] Here, it is then necessary to bring the connection between the modules into line with the possibilities of the simulation environment.

---

[12] Assuming the tie rod is connected with the steering.

Does a superset have to be defined and the unneeded connection points are being deactivated or does a complete set of connection points have to exist per kinematics and these complete sets are being activated or deactivated? How this is implemented depends on the possibilities of the simulation environment. For example, whether there is a macro language available to trigger such a switchover.

3.9.2 Separation of Data and Model

One of the basic paradigms of object-oriented programming is the separation of parameters and model descriptions (source code). This approach should also be taken into account for the modeling of chassis components. Exceptions can at most be very small, non-scalable models in which the effort involved is not worthwhile and which have a more *unique* character. As soon as modular structures are considered, this approach is mandatory, otherwise a reuse of a module is not efficiently possible.

   This means that no concrete values may be entered directly into the model, but that parameters (variables) must be used. These parameters are then loaded into the model with a separate parameter list, usually in the form of a separate file. Using this method, a concrete component can be created from an abstract mechanism (e.g. the suspension kinematics of a trapezoidal-link axle). Two development stages can be compared very quickly, only by exchanging the data set before the respective calculation. For the naming of parameters, clear rules should be developed as discussed in Sect. 6.1.2.

References

[Ammo97]   AMMON, D.: *Modellbildung und Systementwicklung in der Fahrzeugdynamik*, Teubner, Stuttgart, 1997

[HaHo03]   HALFMANN, C. AND HOLZMANN, H.: *Adaptive Modelle für die Kraftfahrzeugdynamik*, Springer, Berlin, 2003

[Ise192]   ISERMANN, R.: *Identifikation dynamischer Systeme, Band 1; Grundlegende Methoden*, Springer, Berlin, 1992

[Ise292]   ISERMANN, R.: *Identifikation dynamischer Systeme, Band 2; Besondere Methoden*, Springer, Berlin, 1992

[KrNe98]   KRAMER, U. AND NECULAU, M.: *Simulationstechnik*, Carl Hanser Verlag, Munich, 1998

[VDI3633] VDI-Richtlinie 3633: *Simulation of systems in materials handling, logistics and production – Terms and definitions.* Verein Deutscher Ingenieure, Duesseldorf, 2018

[Woer11]   WOERNLE, CH.: *Mehrkörpersysteme – Eine Einführung in die Kinematik und Dynamik von Systemen starrer Körper*, Springer-Verlag, Berlin, 2011

Numerical Analysis: The Problem with the Beginning

4

The book of nature is written in the language of mathematics.

Galileo Galilei

4.1   Who Is Euler?

What happens between starting the simulation and saving the results is for many a closed book. Most simulation programs offer the user the possibility to generate various calculation results completely without the knowledge of numerical analysis.

There is usually the option of selecting the integration method and various parameters for calculation. But how can you choose something you do not understand? Common strategies for selecting the integration procedure are:

- Simply use the default settings. *They will have figured something out.*
- Try, until a setting is found where the results are available as quickly as possible. *Time is money.*
- Try, until a setting has been found with as few calculation aborts as possible. *Error-free arithmetic!*
- Or a combination of everything. *Heads or tails?*

You may try as hard as you want with the model, without dealing with the underlying numerical analysis, in the long run, one will only be successful in the rarest cases.

In the next sections, an attempt will be made to support this superficial examination of the subject of numerical analysis from the point of view of an engineer.

© Springer Fachmedien Wiesbaden GmbH, part of Springer Nature 2021   45
D. Adamski, *Simulation in Chassis Technology*,
https://doi.org/10.1007/978-3-658-30678-6_4

The question of who EULER is, will be clarified as well. For those who want or need to dive deeper into this topic, [MuWe12] or [StWP12] are recommended, for example.

4.2   Initial Value Problems or Numerical Integration of Differential Equations

4.2.1   The Initial Value Problem

In the dynamics lecture of technical mechanics, one was confronted with equations of motion, which were formed by differential equations of second order. Let us take as an example a constantly accelerated mass point with constant mass,[1] which after NEWTON's second law produces the following force

$$F = m \cdot \ddot{x} \tag{4.1}$$

Let us convert the equation after the acceleration and name the right side for simplification a_0, so we receive

$$\ddot{x} = \frac{F}{m} = a_0 \tag{4.2}$$

In the lecture, it was always said "through twofold integration we get x". So we will do it the same way here.

$$\dot{x}(t) = a_0 \cdot t + C_1 = v(t) \tag{4.3}$$

$$x(t) = v(t) \cdot t + C_2 = \frac{1}{2} a_0 \cdot t^2 + C_1 \cdot t + C_2 \tag{4.4}$$

Due to the indeterminate integration, we got the two previously unknown integration constants C_1 and C_2 in the Eq. (4.4). Now we come to the keyword *initial value problem*. We can only fill these two constants with life if we know the initial situation of our mass point. As a beginning we designate the point in time $t = 0$ to which we begin our contemplation. Everything that lies before this point in time does not interest us or is implicit in the initial position $x(t = 0) = x_0$ and the initial speed $v(t = 0) = v_0$. As is well known, Eq. (4.3) comes at the point in time of $t = 0$ to $C_1 = v_0$ and with the Eq. (4.4) at the same time to $C_2 = x_0$. Inserted into the two equations, this leads to

---

[1]Those who want to remember what happens when acceleration and mass are not constant, should take a look into their notes of the lecture about dynamics. Additionally, [DaDa13], [Hib312] or [Mayr12] are recommended.

$$\dot{x}(t) = a_0 \cdot t + v_0 = v(t) \tag{4.5}$$

$$x(t) = v(t) \cdot t + x_0 = \frac{1}{2} a_0 \cdot t^2 + v_0 \cdot t + x_0 \tag{4.6}$$

Now we know why it is called the initial value problem. So we always need the condition at the start so that we can calculate everything else. But what is so complicated about integration?

4.2.2 Numerical Integration

The time behavior of technical systems is often described by differential equations. In the general case, however, these are elementary not solvable or the solution of the equations is possible, but very complex. For this reason, the pointwise calculation of the solution by approximation methods is applied. These solution methods are referred to below as integration methods.

What does that mean? The example from the last section can be mentally integrated and in the lectures one usually has functions for the non-constant accelerations, which one could lead to their antiderivatives with simple effort. However, the equations of motion of mechanisms, such as chassis, quickly become complex and can no longer be solved analytically. For this reason, approximation methods are required that estimate the course of the exact function.

In school or university mathematics, everyone has at least encountered the TAYLOR series, with which other functions can be approximated around a work point and which can then be aborted with sufficient accuracy. It will cross our path in this chapter. With the following integration methods, not only position and velocity are to be determined from the acceleration, but also the temporal further development of the acceleration (and thus of the position and velocity).

4.3 Numerical Integration of First Order Differential Equations

For the integration of differential equations one can fall back on a large number of proven algorithms, which are more or less suitable depending on problem definition. The methods used differ, among other things, in the need for information (single or multistep methods), stability (explicit and implicit methods), suitability for special systems (stiff systems) and much more. The most proven methods are usually implemented in mathematical libraries in the programming languages Fortran or C or are offered directly by the simulation programs. Since there is no one method suitable for all applications, a few essential methods are presented in Sect. 4.4 to give an impression of the fundamental differences between the methods.

Fig. 4.1 Single-mass oscillator

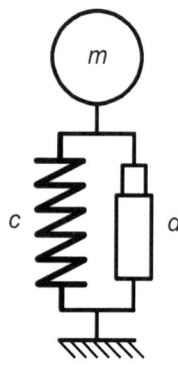

4.3.1 A Simple Example

In order to fundamentally understand the procedures, we use a simple and familiar model—the damped single-mass oscillator (Fig. 4.1). Its behavior is known, its parameters and its equation of motion are quite simple.

The equation of motion for a system without external excitation is described by the following differential equation of second order

$$m\ddot{x} + d\dot{x} + cx = 0 \tag{4.7}$$

Since many of the methods used can only deal with first-order differential equations, this equation must be converted into two first-order differential equations with the substitution

$$
\begin{aligned}
x_1 &= x \\
x_2 &= \dot{x}
\end{aligned}
\tag{4.8}
$$

It follows

$$\dot{x}_1 = x_2$$

$$\dot{x}_2 = -\frac{d}{m}x_2 - \frac{c}{m}x_1 \tag{4.9}$$

The state variable x_1 corresponds to the displacement and x_2 the speed of the mass. The initial condition, for example, can be as follows:

$$\ddot{x} = \begin{pmatrix} x_1 \\ x_2 \end{pmatrix} = \begin{pmatrix} 0 \\ 0 \end{pmatrix} \tag{4.10}$$

At the beginning, the mass is motionless ($v_0 = 0$) in its initial position of deflection ($x_0 = 0$). In order to describe the temporal behavior of the system, we use a numerical integration method.

4.3.2   Polygonal Method According to EULER

First of all: This procedure is an explicit one-step procedure.[2] What exactly this means will be explained later. The task now is to determine the system behavior in the time interval $t_a \leq t \leq t_b$. In the simplest case, the time interval is defined in n steps of the same length

$$h = \frac{t_b - t_a}{n} \tag{4.11}$$

The length h is the so-called step size of the procedure. This length must be chosen appropriately to the time behavior of the system under consideration. This is analogous to the sampling frequency mentioned in Sect. 2.2.2. Since we are dealing with vibrating systems, the sampling frequency should be a multiple of the natural frequency of the system under consideration. If the single-mass oscillator had a natural frequency of 1 Hz, i.e. an oscillation cycle lasts exactly 1 s, a step size of 0.1 s would scan the oscillation ten times. We will see below what this means for the calculation time.

In the last section, the concept of the initial value was discussed, i.e. in the case of our equation of motion, the initial point of time is t_0 the position $x_0 = x(t_0)$ and the speed $v_0 = v(t)_0, x_0)$. The velocity is known to be the first derivative of the position and thus describes the gradient at the given initial position (Fig. 4.2).

Furthermore, we must distinguish between the true function value $x(t_i)$ and the approximate value x_i. If we were to know the course of the function, the exact result would be $x(t_i)$ for the time t_i. Since we do not know it, otherwise we would not have to approach it, we estimate it with x_i.

Starting from the given starting point $P(t_0, x_0)$ which is plotted on the exact solution curve $x = x(t)$ the further course of the function is approximated using the curve tangent.

The calculation rule according to EULER in this case reads

$$x_{i+1} = x_i + h \cdot v(t_i, x_i), \tag{4.12}$$

Where does this rule come from? It is nothing more than a TAYLOR-series (Eq. 4.13), in which the linear element has already been aborted.

---

[2]In the literature, one also finds it under the EULER-CAUCHY method. But, because in most simulation programs only the Swiss mathematician EULER is mentioned, I will focus on him.

Fig. 4.2 Initial values of the function

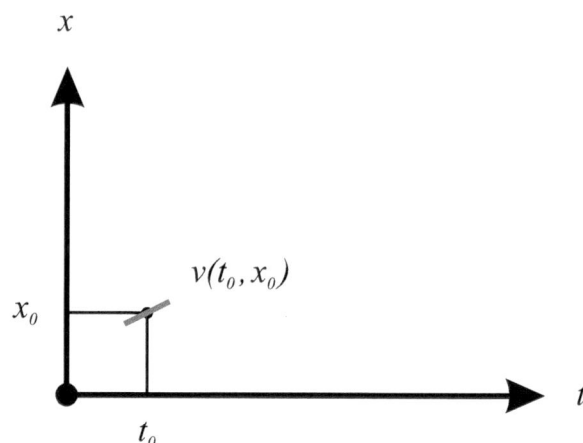

$$f(x) = \sum_{n=0}^{\infty} \frac{f^{(n)}(x_0)}{n!} (x - x_0)^n \tag{4.13}$$

To calculate the next value, we need the next evaluation time. Since in this example we work according to Eq. (4.11) with a constant step size, the next time step is given by

$$t_{i+1} = t_i + h \tag{4.14}$$

Now we use the Eq. (4.12) to the moment t_1 to our problem (Fig. 4.3) and calculate the approximate value x_1.

Also, at this point, we know the gradient again, because we can evaluate the function at this point. It provides us with the first derivation and thus the gradient.

The equation solution method, which can look different in each simulation program, converts the equations of motion into Eq. (4.9) so that the derivatives are available for integration.

We repeat this procedure until we have reached the end time of the observation (Fig. 4.4).

The described integration procedure therefore allows us to take a look into the future. In this way, we can estimate how the system will behave approximately over time.

4.3.3   Types of Errors

4.3.3.1 Local Error

Since this is an approximate calculation, errors are made. A distinction is made between the local and the global error. The local error e_i which we have at the time t_i is defined as follows

Fig. 4.3 Further development
of the function

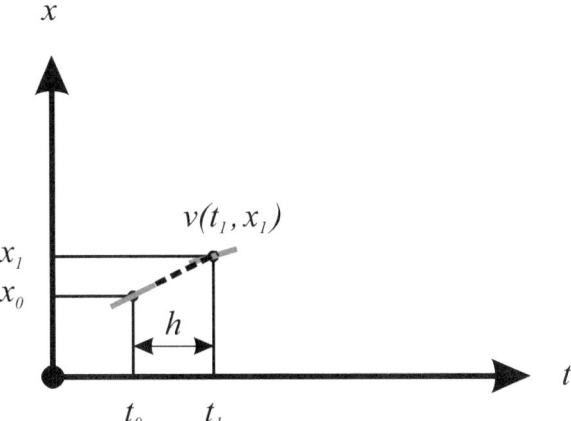

Fig. 4.4 End value of the
function

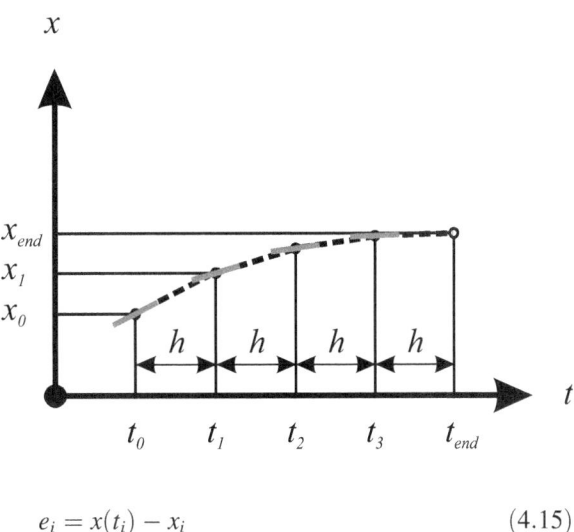

$$e_i = x(t_i) - x_i \tag{4.15}$$

It is the difference between the true function value and the approximate value at the time. It is the error of the procedure which, assuming that the last value was exact, is caused by the approximation.

4.3.3.2 Global Error

If several calculations with errors are executed one after the other, a global error is generated over the entire period, as in Fig. 4.5 for the time point t_{end} is depicted. This is because each calculated value is already based on an incorrect value. The error reproduces itself. Ultimately, the global error is of interest because it shows the total deviation between the calculated value and the expected value.

Fig. 4.5 Global error

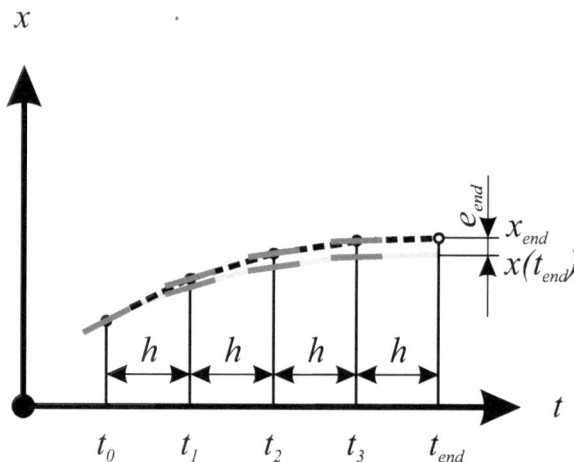

But how can you calculate the error if you do not know the exact value of the function? You can at least talk about the TAYLOR-series approach to determine the order of the error.[3] According to this, statements such as "the local error e_i goes with h^2 close to zero if h goes to zero" is possible.

In the literature, one often finds the spelling with the LANDAU symbol O in the form $e_i = O(h^2)$. This means that by halving the step size, the local error is reduced to a quarter. Now, of course, you will want to make the step size as small as possible to minimize the error. What in this case is meant by *possible*, we will discuss it later.

4.3.3.3 Rounding Error

Section 2.2.2 already discussed the discretization of numerical values and that the accuracy of representation in a digital computer is limited. The concrete meaning for numerical integration is that for particularly small numbers the inaccuracy increases due to rounding. Since the calculation then continues with rounded numbers, this error also reproduces itself. The minimum step size is thus limited by the accuracy of the representation of the floating point numbers in the computer used and, as we will see below, by the patience of the computational engineer.

4.3.3.4 Total Error

As Fig. 4.6 shows, one has to find a compromise between the rounding error and the process error (global error) in order to choose the correct step size. Another not insignificant point is the calculation time. If we stick with the previous integration method, we have

---

[3]At this point, I refrain from giving the derivation. The books about numerical analysis are full of them, however, one should at least understand what it means when a growth of errors is described with $O(h^2)$.

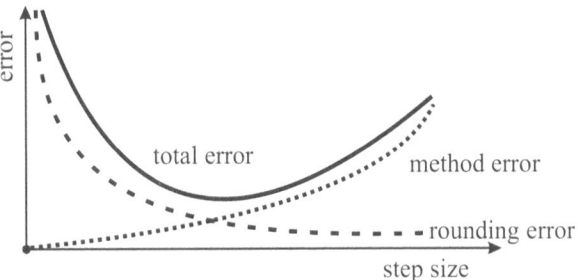

Fig. 4.6 Total error from the global error and the rounding error

to evaluate the equations of motion once for each calculation step. If the simulation time is to be 10 s, we must evaluate the equations of motion 10,000 times with a step width of 1 ms. If we reduce the step size to 0.1 ms, we have to evaluate the equations of motion 100,000 times and so on. In this example, the calculation time is set with $h = 0.1$ ms ten times longer than at the beginning with $h = 1$ ms. The integration methods presented in Sect. 4.4 evaluate the equations of motion per calculation step several times, so that the effort increases accordingly. However, these procedures are usually equipped with a step size and/or order control, so that in the end a computing time advantage can result.

If you are working with a new model class, a new integration procedure or with new load cases, you should start by performing the same calculation with different error limits. The goal is that the solution, i.e. the result of the calculation, converges, i.e. strives towards a fixed value (more details in the next section). In the case of the calculations, the results should no longer change when the error limits become more accurate above a certain level of accuracy. A refinement does not lead to an improvement of the result—only an increase of the computing time. This way it is possible to find out which settings are sufficiently accurate for the respective combination of model, integration method and load case.

If the model has discontinuities (e.g. due to steps in characteristic curves) or in the case of a load, sudden forces are introduced (e.g. during obstacle crossings), this can lead to increased requirements on the accuracy of the calculation procedure.

4.3.4   Convergence and Stability

The first task is to clarify what the concepts of convergence and stability mean in numerical analysis. In the community of the computational engineers the colloquially established term is *runs stable* if the calculation runs without interruption.[4] To say it like a mathematician: "This is necessary, but not sufficient."

---

[4]In this context, the term flawless is often mentioned too. However, this term only describes the absence of error reports from the simulation program and has nothing to do with the erroneous discussed in Chap. 3.

4.3.4.1 Convergence

Convergence is the principle of hope of the computational engineer. A numerically approximated system converges when it strives towards the exact solution with decreasing step size. In the borderline case, when the step size is close to zero, the approximation thus represents the exact system behavior. Since the step sizes used are small, but not equal to zero, the exact value is not reached. A residual error remains. In everyday computing, the statement that the system converges means that it remains within a specified error limit. The most feared error message of the integration procedure is that no solution could be found and the calculation had to be aborted. Conclusion: This combination of system behavior and integration methods does not converge.

When this message occurs, there are different reactions depending on the experience and time pressure of the computational engineer. The unreflected testing of all offered integration methods is popular until the error message disappears. Explicit procedures without step size or order control do not report convergence problems because they lack the measure of comparison. For this, you can then use the *error-free* calculation in the animation of the system or in the time series to consider an unexpected system behavior.

Another possibility is the reduction of error barriers. You do not take it so seriously anymore and leave the integration process a margin. The questions that arise are:

1. Why was the value smaller before?
2. Was it even necessarily that small?

If not, you have invested a lot of time in a supposed accuracy. If it is, why should inaccurate behavior be accepted now?

There is no getting around taking a closer look at your system and finding the causes of the lack of convergence. One result, of course, may be that the chosen integration method does not actually fit this class of systems or load cases. Or one finds a wrong or insufficient modeling and has to improve it.

4.3.4.2 Concept of Stability

In numerical analysis and systems engineering, a system is considered stable when all influences of disturbances and errors subside over time. The system behavior asymptotically approaches a stationary state, if new disturbances are not added. In the case of the integration procedures discussed here, this means that the error assumes a constant value. With unstable systems, the error would increase over time. If the error oscillates around a constant value (continuous oscillation), this is referred to as limit stability.

Since we are dealing with models of real systems, which we have to consider numerically, the statement applies that a system cannot be described as stable or unstable on the basis of its model definition alone. Rather, a physically stable system may appear unstable due to inappropriate numerical treatment and a physically unstable system may appear stable. This effect was described in [RiSc10] as so-called numerical damping as follows.

4.3.4.3 How Stable Systems Become Unstable

From dynamics, the undamped single-mass oscillator is known, whose equation of motion without external excitation is as follows

$$m\,\ddot{x}(t) + c\,x(t) = 0 \tag{4.16}$$

In vibration theory, this equation is used to determine the natural frequency ω_0 as follows

$$\ddot{x}(t) + \frac{c}{m}\,x(t) = \ddot{x}(t) + \omega_0^2\,x(t) = 0 \tag{4.17}$$

So, for the acceleration and for the stiffness follows

$$\ddot{x}(t) = -\omega_0^2\,x(t) \tag{4.18}$$

$$c = m\,\omega_0^2 \tag{4.19}$$

For the undamped single-mass oscillator, assuming that this is a conservative system, the sum of potential and kinetic energy is always constant. The spring serves as energy storage, the dissipating damper is missing. This is the principle of conservation of energy

$$E = \frac{1}{2}m\,\dot{x}2 + \frac{1}{2}c\,x^2 = \frac{1}{2}m\left(\dot{x}^2 + \omega_0^2\,x^2\right) = const \tag{4.20}$$

Let us consider the temporal behavior using the previously described explicit EULER's method, then for the motion quantities the Eq. (4.18) gives

$$x(t + h) = x(t) + h \cdot \dot{x}(t) \tag{4.21}$$

$$\dot{x}(t + h) = \dot{x}(t) - h \cdot \omega_0^2\,x(t) \tag{4.22}$$

If we put this in Eq. (4.20), we get

$$E(t + h) = \frac{1}{2}m\left[\left(\dot{x}(t) - h \cdot \omega_0^2\,x(t)\right)^2 + \omega_0^2(x(t) + h \cdot \dot{x}(t))^2\right]$$

$$= \frac{1}{2}m\left[\dot{x}^2(t) - 2\,\dot{x}(t)\,h\,\omega_0^2\,x(t) + h^2\,\omega_0^4\,x^2(t) + \omega_0^2\left(x^2(t) + 2\,x(t)h\,\dot{x}(t) + h^2\,\dot{x}^2(t)\right)\right]$$

$$= \frac{1}{2}m \left[\left(1 + h^2\omega_0^2\right) \dot{x}^2(t) + \left(1 + h^2\omega_0^2\right) \omega_0^2 x^2(t) \right]$$

$$= \left(1 + h^2\omega_0^2\right) \frac{1}{2}m \left(\dot{x}^2(t) + \omega_0^2 x^2(t)\right)$$

$$= \left(1 + h^2\omega_0^2\right) E(t) \tag{4.23}$$

The last line is to be understood as a contradiction to the principle of conservation of energy, since the energy must be constant at all times. It increases here. This will cause the system to oscillate, although no further energy will be added from the outside. It becomes unstable—through pure *numerical energy*.

4.3.4.4 How Unstable Systems Become Stable

For the same single mass oscillator as in the previous section, the implicit EULER's method is described in more detail in Sect. 4.4.2.

For the basic Eqs. (4.21) and (4.22), this means that

$$x(t + h) = x(t) + h \cdot \dot{x}(t + h) \tag{4.24}$$

$$\dot{x}(t + h) = \dot{x}(t) - h \cdot \omega_0^2 \, x(t + h) \tag{4.25}$$

This results in the following form according to the same procedure as in the last section without proof for the working principle

$$E(t + h) = \frac{1}{1 + h^2\omega_0^2} E(t) \tag{4.26}$$

This in turn is a violation of energy conservation, because the energy decreases with each calculation step. This behavior corresponds to a damped system and is stable. However, since there is no damper physically present at all, in this case it is a matter of *numerical damping*. In [RiSc10] this case is classified as more favorable, since in the reality of technical mechanics there are no undamped systems. This way, unmodeled damping effects such as neglected friction or wind resistance would be taken into account. I can follow that view to the extent that the user knows about it.

Of course, there are also integration methods that would satisfy the principle of energy conservation for our example. At this point, however, the hint that this effect exists and should be observed should suffice. We do not want to leave the surface of numerical analysis.

Therefore, the stability of the system to be simulated as well as the stability of the numerical method used must be kept in mind.

4.4   Integration Methods

Without claiming to be exhaustive, a few of the most important and best-known integration methods are presented below. In this way, a basic understanding of the procedures that take place in the background of the calculation is to be conveyed.

4.4.1   Method Overview

4.4.1.1 One-Step and Multistep Methods

One-step methods only use the current time for the calculation of the next time step. The previous knowledge about the values already calculated is not used (Fig. 4.7).

The already presented EULER's method (Sect. 4.3.2) and the later described method of RUNGE-KUTTA are one-step procedures. However, the calculation of the equations of motion in the latter case requires several evaluations. Nevertheless, one speaks of a one-step procedure.

The multistep procedures have a memory, i.e. they remember previous steps and use them to calculate the new step (Fig. 4.8).

In this way, evaluations of the equations of motion can be saved by using suitable polynomial approaches. These require a lot of computing time, so that the multistep procedures usually have a speed advantage.

The ADAMS- and BDF methods, as discussed in Sects. 4.4.4 and 4.4.5, are multistep methods. You use the knowledge of the previous function progression and use it to predict the further progression. For the first points in time to be calculated, there are not yet enough historical values available, which is why some methods start with one-step methods until the necessary number of sampling points has been reached. The complexity of the process then gradually increases.

4.4.1.2 Explicit and Implicit Methods

A further distinction between the methods is whether it already uses the value to be calculated or only uses values from the past. This is independent of whether the method is one-step or multistep.

The previously treated EULER's method comes with the addition *explicit* as it is based only on the already known state. For the calculation of the approximate value, only values which lie before the approximate value to be determined are used. The number of required calculation steps for the calculation of the approximate value is known a priori and thus also the calculation time requirement.

From this it can be concluded that the implicit methods also use the approximate value itself for the calculation of the approximate values. Since the value is of course not yet available, an iterative correction procedure must help, as described in the following section. However, this means that the required calculation steps are dependent on error limits and the current system behavior. The calculation time of a calculation step can therefore vary.

Fig. 4.7 One-step method

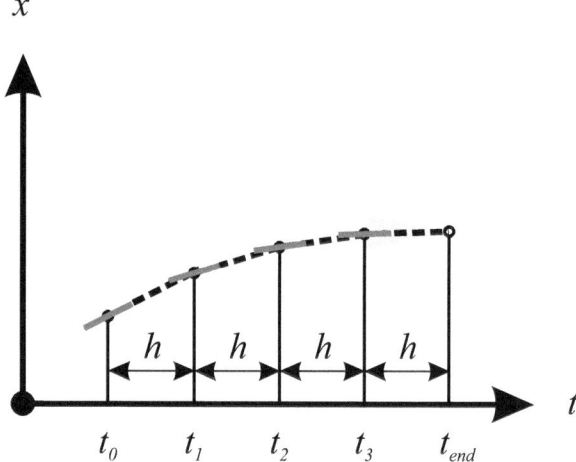

Fig. 4.8 Multistep method

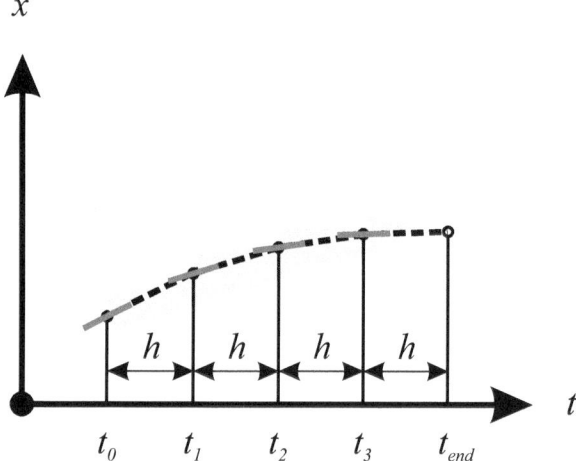

This is not possible for every application, because hardware-in-the-loop simulations must always comply with the real-time condition (Sect. 17.1).

4.4.2 Implicit EULER's Method

The implicit EULER's predictor-corrector method consists of two steps, because the so-called predictor-corrector technique is used here. First, the predictor step is an explicit EULER-step (see Eq. 4.12).

Fig. 4.9 Predictor step

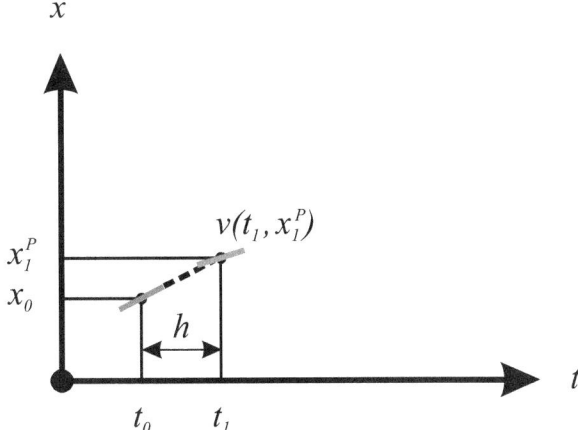

$$x_{i+1}^{P} = x_i + h \cdot v(t_i, x_i) \tag{4.27}$$

Now we have, as can be seen in Fig. 4.9, a provisional approximate value. At this point we can now evaluate the equations of motion and obtain the slope at this point, which we can use in Eq. (4.28).

Next, the corrector step is performed at this newly calculated position (Fig. 4.10).

$$x_{i+1}^{K} = x_i + h \cdot v\left(t_{i+1}, x_{i+1}^{P}\right) \tag{4.28}$$

In the end, the gradient of the estimated point is taken and applied to the starting point. Changes in the gradient are therefore taken into account at an early stage. However, in this example, the equations of motion must be called twice instead of once as in the explicit procedure.

A variant is the procedure of HEUN the bisector of the angle of the two pitches is used. The calculated value then lies between the value of the explicit and implicit EULER's method.

Higher procedures use error or convergence estimates and repeat the two steps with different step sizes until they fall below the specified error limits or exceed the maximum number of iteration steps. The latter leads to the cancellation of the calculation because no solution was found.

I remind you once again that v is the derivation of x which is calculated by the function call of the equations of motion.

In the literature on numerical methods, EULER's method is often described as too imprecise and therefore insufficient for practical use. Practice shows something different. Due to the speed and simplicity of the procedure, it is particularly popular for hardware-in-the-loop and vehicle dynamics simulations. Especially the inexperienced user is pleased

Fig. 4.10 Corrector step

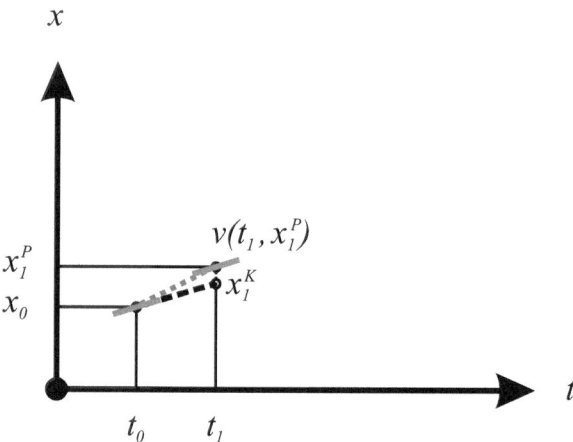

that he only has to enter one parameter, the step size.[5] For the consideration of the linear part of vehicle dynamics, this procedure seems to me to be completely sufficient. At the transition into the nonlinear part, especially in the border area, one has to take a closer look to see if the real system is still sufficiently accurately mapped.

4.4.3   RUNGE-KUTTA Method

Based on the EULER's method, the RUNGE-KUTTA-method uses more information. The procedure of the fourth order is composed in its so-called classical variant[6] of the following steps that originate from the TAYLOR's series development:

$$x_{i+1} = x_i + \frac{1}{6}(k_1 + 2\,k_2 + 2\,k_3 + k_4) \tag{4.29}$$

with the coefficients

$$k_1 = h \cdot v\,(t_i, x_i) \tag{4.30}$$

---

[5]The most popular step size will probably be 1 ms. Because the system behavior is classically measured to up to 5 Hz in vehicle dynamics, this leads to at least 200 measuring points per oscillation cycle.

[6]Further variants exist in which the weighting factors are chosen differently.

$$k_2 = h \cdot v \left(t_i + \frac{h}{2}, x_i + \frac{k_1}{2} \right) \tag{4.31}$$

$$k_3 = h \cdot v \left(t_i + \frac{h}{2}, x_i + \frac{k_2}{2} \right) \tag{4.32}$$

$$k_4 = h \cdot v \left(t_i + h, x_i + k_3 \right) \tag{4.33}$$

With the explicit EULER's method the gradient at the left edge of the interval is used exclusively. In the RUNGE-KUTTA method of the fourth order this corresponds to the coefficient k_1 which flows in at one-sixth. In addition, two gradients are in the interval center, k_2 and k_3 to one third each, and the gradient at the end of the interval, k_4 with a sixth. This results in a mean gradient which, due to the weighting of the grid points, should better represent the further course of the function (Fig. 4.11). Since no values from the last calculation interval and the function value to be calculated are used, this is an explicit one-step method.

The increased accuracy has its price, because four function calls are necessary. This can only be compensated due to the higher error order and the resulting larger step size.

4.4.4 ADAMS Method

The previously described RUNGE-KUTTA-method uses four grid points within an interval to approximate a new function value. The grid points are used to better estimate the local course of the function within the next interval.[7] The multistep methods, on the other hand, use the function values already calculated in the past for estimating the function course, so that several intervals provide the corresponding grid points. The best known methods are the two ADAMS-methods. With the ADAMS-BASHFORTH method, there is an explicit procedure (Fig. 4.12) and with the ADAMS-MOULTON method, an implicit procedure available. Both are linear methods that use the superposition of the individual results.

It is essential for the method that the coefficients of an n-order interpolation polynomial are found. The higher the order of the procedure, the more function evaluations must be used, but the higher the order of the errors. A major advantage over the higher-order one-step methods is that only one function evaluation for the explicit and two function evaluations for the implicit methods must be performed per calculation step. All other function evaluations have already been calculated for the previous steps—they only have to be carried out in the *remembrance*.

---

[7]Predetermined by the step size h.

Fig. 4.11 RUNGE-KUTTA-fourth order method

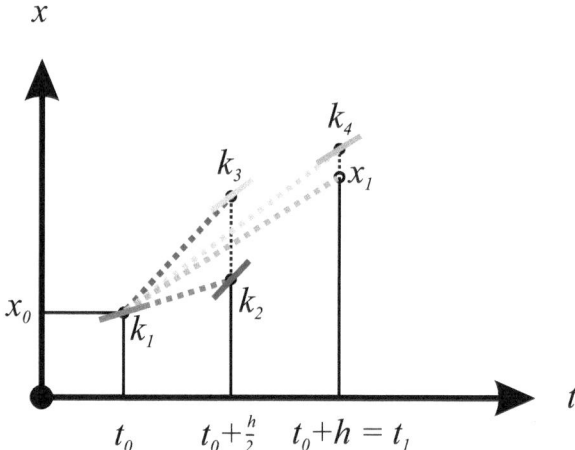

Fig. 4.12 ADAMS-BASHFORTH method

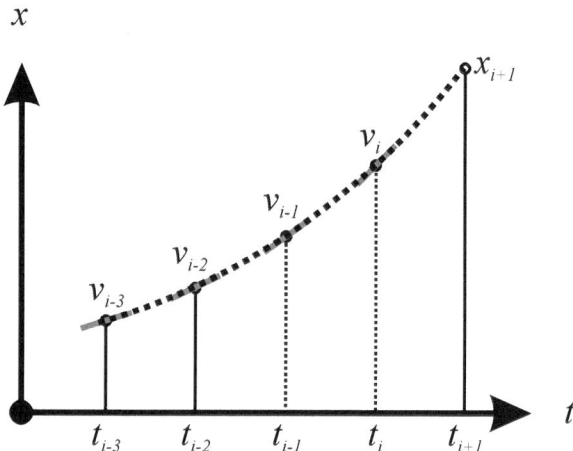

   With the explicit method, the order of errors corresponds to the order of the method, with the implicit method it is in each case one order higher than that of the method. For the ADAMS-BASHFORTH-method, this is shown as an example in Table 4.1.

4.4.5   BDF Method

If the natural frequencies of the subsystems differ greatly within a system, one speaks of a *stiff* differential equation. As a guide, one can take a ratio of the largest to the smallest natural frequency of at least 1000. If, in addition to the rigid bodies commonly used for multi-body systems, elastic structures are also used, which are represented, for example, by

Table 4.1 Calculation rules of the ADAMS-BASHFORTH method

Order	Method	Error order
1[a]	$x_{i+1} = x_i + h \cdot v\,(t_i, x_i) = x_i + h \cdot v_i$	1
2	$x_{i+1} = x_i + \frac{h}{2} \cdot (3\,v_i - v_{i-1})$	2
3	$x_{i+1} = x_i + \frac{h}{12} \cdot (23\,v_i - 16\,v_{i-1} + 5\,v_{i-2})$	3
4	$x_{i+1} = x_i + \frac{h}{24} \cdot (55\,v_i - 59\,v_{i-1} + 37\,v_{i-2} - 9\,v_{i-3})$	4

[a]Corresponds with the explicit EULER method

beam models, or if the hydraulics of the brake system are also represented, large differences in the natural frequencies can quickly occur.

The integration methods discussed so far would have to orient themselves with their step size or their order to the largest natural frequency and accordingly use very small integration step sizes. The consequences have already been discussed. The calculation time increases strongly, the rounding errors increase and depending on the integration method, the calculation is terminated under circumstances, since no convergence is reached.

Implicit methods are better suited to such problems than explicit ones. A special class of integration methods has been developed for this type of system. The so-called *Backwards Differential Functions* or better yet, by its abbreviation *BDF* are also linear multistep methods. In contrast to the ADAMS-MOULTON-method, they have a larger stability area up to order six and can therefore allow larger step widths. With *BDF* method integrals are not approximated by an interpolation polynomial, but by the derivative v_{i+1}.

4.5   Interpolation and Extrapolation Methods

4.5.1   Interpolation

Measured component characteristic curves are available in the form of discrete measuring points which represent the grid points of the characteristic curves. During the simulation also values are needed, which can lie arbitrarily between these points. For this reason, the corresponding values must be interpolated.

Various methods are often offered for interpolation. Linear interpolation is usually used as the standard method. It is very fast because once the correct range of values has been determined, it only has to determine the gradient between two points (Fig. 4.13).

The value $P(x, y)$ that lies between the values $P(x_i, y_i)$ and $P(x_{i+1}, y_{i+1})$ (Fig. 4.14), is determined according to the following regulation:

$$y = \frac{x - x_i}{x_{i+1} - x_i} \cdot (y_{i+1} - y_i) + y_i \tag{4.34}$$

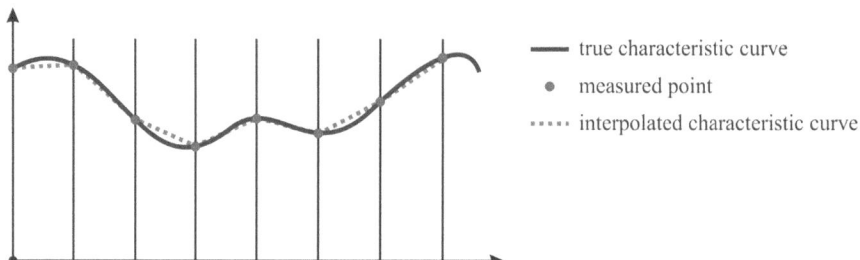

Fig. 4.13 Linear interpolation

Fig. 4.14 Determination of the
value to be interpolated

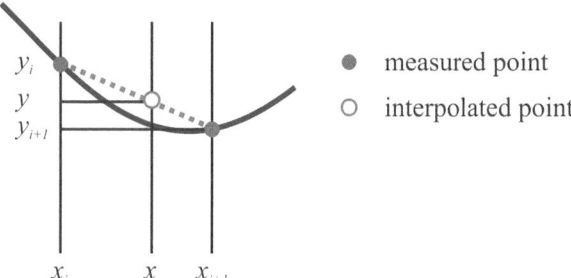

The clear disadvantage of this method comes to bear when the gradient changes sharply, resulting in discontinuous transitions that can lead to numerical problems. The surroundings of the two points is not taken into account, so that discontinuities cannot be excluded by the linear interpolation method.

Methods that avoid these discontinuities, but require more computational effort, rely on higher polynomial approaches. A common method is to place a spline function through the grid points. A spline consists of polynomials which represent only a part of the curve and guarantee a continuous course at the transitions.

A frequently offered variant is the cubic spline, to which a third-order polynomial is assigned.

$$f(x) = a_i x^3 + b_i x^2 + c_i x + d_i \tag{4.35}$$

For this purpose, suitable coefficients a_i, b_i, c_i and d_i are used. so that the function can be steadily differentiated twice. The resulting curves always look *smooth* which simulates a continuous process for the intermediate values that may not correspond with the modeled system.

It is helpful, if the simulation software can display the characteristic curve with the selected interpolation method, because unwanted results can be achieved. As can be seen in Fig. 4.15, the few grid points provided do not lead to a clear solution. Both the cubic spline

Fig. 4.15 Effects of different interpolation methods

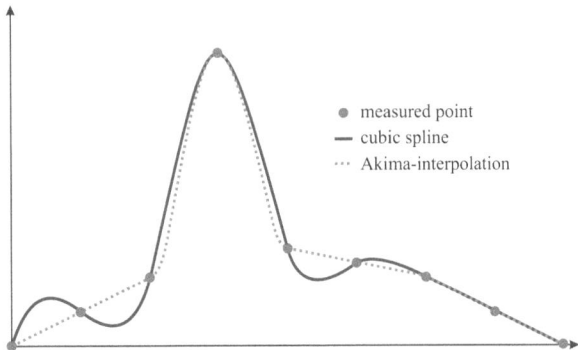

- measured point
- cubic spline
- Akima-interpolation

and the Akima approach[8] could show the desired behavior. But as you can clearly see, it does not lead to the same result if you use the procedure *by accident*. A larger number of grid points would lead to a convergence of procedures. The example shows drastically that an optical control of the interpolation result is very important.

If the zero crossing must be hit exactly, it must, measured or not, usually be specified explicitly in the characteristic curve, since otherwise the interpolation procedure may calculate a value other than zero due to the two surrounding interpolation points.

4.5.2   Extrapolation

The measured characteristic curve is available in a certain value range, which is often defined by the limits of the measuring device or procedure. In the rarest cases it can be predicted with certainty that no values outside this measuring range have to be evaluated within the scope of the simulation. In this case, most simulation tools provide for extrapolation beyond the limits of the measurement. The same procedures are often offered for extrapolation as for the interpolation described above.

It should be noted that the last points, often only the last two in linear extrapolation, are used for the calculation. Figure 4.16 shows the influence of an intermediate point on the expected gradient. Where at the end of the characteristic curve strong gradient changes are present—this is the case with progressive characteristic curves, in particular with bump stop characteristic curves—particular importance must be placed on the quality of these sampling points. Otherwise highly implausible values can be determined. In the case of bump stops, this can lead to exorbitantly high forces, which in reality would lead to the destruction or at least deformation of components. In the rigid body world of multi-body systems, nothing breaks down so quickly, but the large forces can then lead to unusual movements in the mechanism.

---

[8]In this method, cubic splines are used too. However, the neighboring intervals are considered in the calculation of the derivatives at the intervals' transitions.

Fig. 4.16 Influence of endpoints on extrapolation

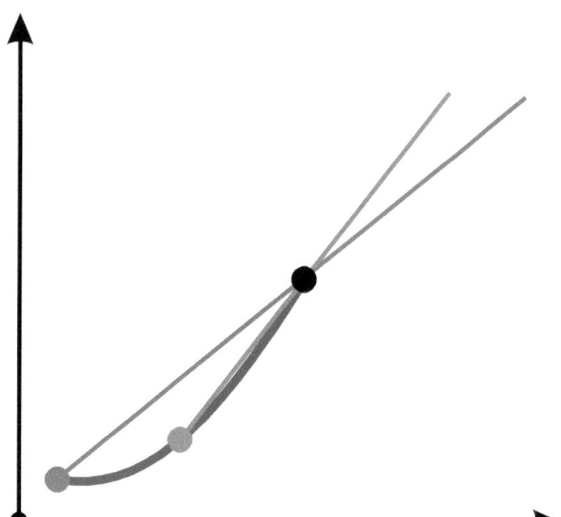

4.6   Functions for Fading in and Fading out

From my lectures I know that many students—due to a lack of experience—have problems finding suitable approaches and parameters for fading in or fading out functions and signals. Since you never know who is holding this book in their hands at the moment, I would like to take this opportunity to think of those who have similar difficulties. Everyone else can lean back and skip this section.

The task is to convert a given signal or function from a start value to a final value within a given period of time. In the following, I will introduce three common procedures for scaling from 0% to 100% or vice versa. Any other function context and any other scaling can also be used. An adaptation of the described procedures should not be difficult.

4.6.1   Linear

The easiest way to get from a start value to a target value is to use the straight line equation $y = m\,x + b$. It is well known that two points are sufficient to describe a straight line unambiguously. Figure 4.17 shows two possible courses.

As a scaling factor from 0% to 100% (left) or from 100% to 0% (right), these linear equations can be used in the displayed time range of 4 s. The factor k regulates the slope of the straight line, i.e. in this case the time after which the value one is reached. Even if the equation parameters are already recognized by looking sharply, the formal calculation for the left straight line in Fig. 4.17 follows. The straight line equation has two unknowns and we have two coordinates, so this is *solvable*.

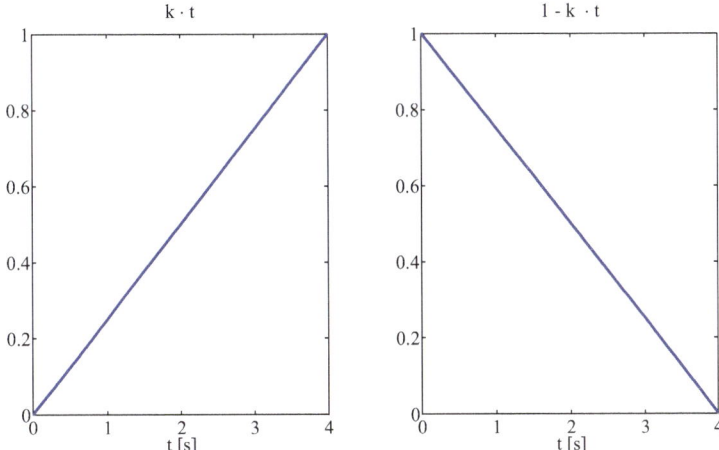

Fig. 4.17 Scaling with the straight line equation

$$P_1(0,0) : m \cdot 0 + b = 0 \rightarrow b = 0 \tag{4.36}$$

$$P_2(4,1) : m \cdot 4 + 0 = 1 \rightarrow m = \frac{1}{4} \tag{4.37}$$

Such scaling is often used to prevent a signal from starting directly with a value unequal to zero, which could lead to larger jumps when controlling a test bench, for example.

For times $t > 4$ s the scaling must of course not increase further, but must be limited to one (Eq. 4.38) or zero (Eq. 4.39) by the software. As usual, not everything that is mathematically possible also makes physical sense.

$$y(t) = \min\left(\frac{t}{t_E}; 1\right) \cdot f(t) \tag{4.38}$$

$$y(t) = \max\left(1 - \frac{t}{t_E}; 0\right) \cdot f(t) \tag{4.39}$$

4.6.2   Exponential

If the time behavior is to be integrated into the model, an exponential rise or fall of the value is often used. First, let us look at the use of the e-function in the progressive or degressive course of the rise or fall (Fig. 4.18).

Depending on the desired course, the basic function must first be selected. In the following examples it is always assumed that target values are available for two points in

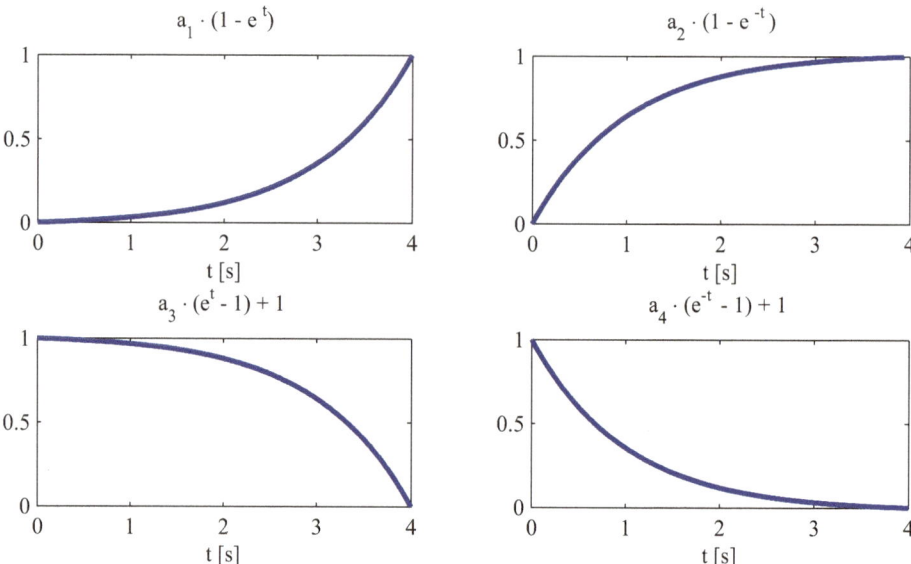

Fig. 4.18 Scaling with the e-function

time that are to be achieved. The solutions presented below are of course only examples. There are a variety of solutions, especially if the respective gradients at the known points are to be taken into account. The values of the example are the same as for the straight line equation. For the limitation of the scaling function, the same correlations apply as for the straight line equation.

4.6.2.1 Progressive Increase

The upper left graph shows a progressive course, which begins at $t = 0$ s with the value zero and ends at $t_E = 4$ s with the value one. The function

$$y_1 = a_1 \cdot (1 - e^t) \tag{4.40}$$

has been chosen, so that the following equations can be evaluated

$$P_1(0,0) : a_1 \cdot (1 - 1) = 0 \tag{4.41}$$

$$P_2(4,1) : a_1 \cdot \left(1 - e^4\right) = 1 \rightarrow a_1 = \frac{1}{1 - e^4} \tag{4.42}$$

Used in Eq. (4.40) the following results

$$y_1 = \frac{1 - e^t}{1 - e^4} = \frac{1 - e^t}{1 - e^{t_E}} \qquad (4.43)$$

4.6.2.2 Degressive Increase

The upper right graph shows a degressive course, which begins at $t = 0$ s with the value zero and ends at $t_E = 4$ s with the value one. The function

$$y_2 = a_2 \cdot (1 - e^{-t}) \qquad (4.44)$$

has been chosen, so that the following equations can be evaluated

$$P_1(0,0) : a_2 \cdot (1 - 1) = 0 \qquad (4.45)$$

$$P_2(4,1) : a_2 \cdot (1 - e^{-4}) = 1 \rightarrow a_2 = \frac{1}{1 - e^{-4}} \qquad (4.46)$$

Used in Eq. (4.44) the following results

$$y_2 = \frac{1 - e^{-t}}{1 - e^{-4}} = \frac{1 - e^{-t}}{1 - e^{-t_E}} \qquad (4.47)$$

4.6.2.3 Progressive Decrease

The lower left graph shows a progressive course, which begins at $t = 0$ s with the value one and ends at $t_E = 4$ s with the value zero. The function

$$y_3 = a_3 \cdot (e^t - 1) + 1 \qquad (4.48)$$

has been chosen, so that the following equations can be evaluated

$$P_1(0,1) : a_3 \cdot (1 - 1) + 1 = 1 \qquad (4.49)$$

$$P_2(4,0) : a_3 \cdot (e^4 - 1) + 1 = 0 \rightarrow a_3 = \frac{1}{1 - e^4} \qquad (4.50)$$

The coefficient a_3 corresponds to that of a_1. For Eq. (4.48) this means

$$y_3 = 1 - \frac{1 - e^t}{1 - e^4} = 1 - \frac{1 - e^t}{1 - e^{t_E}} \qquad (4.51)$$

4.6.2.4 Degressive Decrease

The lower right graph shows a progressive course, which begins at $t = 0$ s with the value one and ends at $t_E = 4$ s with the value zero. The function

$$y_4 = a_4 \cdot (e^{-t} - 1) + 1 \tag{4.52}$$

has been chosen, so that the following equations can be evaluated

$$P_1(0, 1) : a_4 \cdot (1 - 1) + 1 = 1 \tag{4.53}$$

$$P_2(4, 0) : a_4 \cdot (e^{-4} - 1) + 1 = 0 \rightarrow a_4 = \frac{1}{1 - e^{-4}} \tag{4.54}$$

Here, too, the coefficient corresponds to a_4 that of a_2. For Eq. (4.52) this means

$$y_4 = 1 - \frac{1 - e^{-t}}{1 - e^{-4}} = 1 - \frac{1 - e^{-t}}{1 - e^{-t_E}} \tag{4.55}$$

4.6.3   Trigonometric

As is clearly shown in Fig. 4.17 for the straight line equation and in Fig. 4.18 for the e-function, the gradients at the end points are not zero, so there will be a discontinuous transition for times greater than the 4 s selected here. This is not acceptable for all applications. Let us start from an example where two signals are to be connected to each other and both have a gradient of zero at the transition point.[9]

Of course for the e-function *more complex* equations are found whose derivation can be zero, but there is also another elegant method. We use the cosine,[10] which is located at the points $P(0, 1)$ and $P(\pi, 1)$ has a gradient of zero (Fig. 4.19). Here, too, there are two variants for fading in and fading out the signal (Fig. 4.20). Since the cosine is an angle function which runs periodically between the angles 0 and 2π, we first have to convert it into the time domain and shorten it to half, since we only need the branch up to π.

The constant ω results from the requirement of our standard example that half of the period should expire in 4 s.

---

[9]Of course the values are unequal zero, otherwise they could be directly joined together.

[10]Since the sinus is just a shifted cosine, it will not be mentioned any further here. However, it can be used without worry by the supporters of this function.

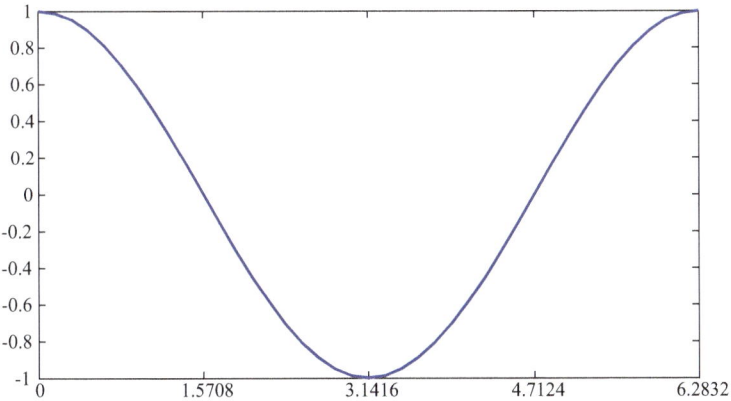

Fig. 4.19 Cosine function

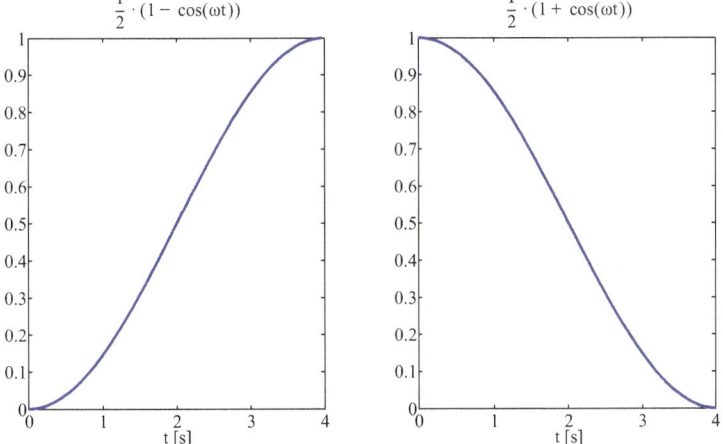

Fig. 4.20 Scaling with the cosine function

$$\omega = \frac{\pi}{4\ s} = \frac{\pi}{t_E} \tag{4.56}$$

Since the cosine oscillates between its maxima $+1$ and -1, the amplitude must be halved and shifted by an offset so that the start and end values are reached. This way, the two scaling functions

$$y_{0\rightarrow1} = \frac{1}{2}\left(1 - \cos\left(\omega t\right)\right) \tag{4.57}$$

$$y_{1 \to 0} = \frac{1}{2} \left(1 + \cos\left(\omega t\right)\right) \tag{4.58}$$

are made. The above applies to the limitation of the function to the period from 0 to 4 s.

References

[DaDa13]   DANKERT, J. AND DANKERT H.: *Technische Mechanik: Statik, Festigkeitslehre, Kinematik/Kinetik*, Springer, Berlin, 2013

[Hib312]   HIBBELER, R.C.: *Technische Mechanik 3 – Dynamik*, Pearson Studium, Munich, 2012

[Mayr12]   MAYR, M.: *Technische Mechanik*, Hanser Verlag, Munich, 2012

[MuWe12]   MUNZ, C. D. AND WESTERMANN, TH.: *Numerische Behandlung gewöhnlicher und partieller Differenzialgleichungen*, Springer, Heidelberg, 2012

[RiSc10]   RILL, G. AND SCHAEFFER, TH.: *Grundlagen und Methodik der Mehrkörpersimulation*, Vieweg+Teubner, Wiesbaden, 2010

[StWP12]   STREHMEL, K., WEINER, R. AND PODHAISKY, H.: *Numerik gewöhnlicher Differentialgleichungen – Nichtsteife, steife und differential-algebraische Gleichungen*, Springer, Wiesbaden, 2012

Simulation Tools

5

5.1 Tool Selection

Should the reader be embarrassed at a later point in time that a tool is to be selected for the planned tasks or that an existing one is to be replaced by a new one, then the following is to be considered to him as *decision-making basis* to the hand. If you do not yet have a simulation environment, you first have the choice between *make or buy,* i.e. the self-created in-house solution (Sect. 5.1.1) or the purchase of a commercial product (Sect. 5.1.2).

5.1.1 In-House Solution

The clear advantage of an in-house solution is its adaptability to one's own ideas and needs. Since you program the simulation environment yourself, you can optimally adapt it to your own process and know what happens below the user interface.

The clear disadvantage of this approach is usually answered by the question whether the programming of such a tool belongs to one's own core competence and whether one has the skills to do it at all. Normally, the question of core competence outside universities and research institutions will be answered negatively. The necessary capabilities to create a simulation environment should also be available in very few cases. Furthermore, it should be borne in mind that the subsequent maintenance of the program itself must also be guaranteed in the event of personnel changes.

© Springer Fachmedien Wiesbaden GmbH, part of Springer Nature 2021
D. Adamski, *Simulation in Chassis Technology*,
https://doi.org/10.1007/978-3-658-30678-6_5

Furthermore, it has to be clearly distinguished here whether one is only using the *small* tool for one's own use and for a limited number of applications or whether the use is also intended for third parties and over a longer period of time. The programming effort increases immensely, because these *third parties* do not necessarily know what this tool is all about and what input and data formats are expected. So you have to do more documentation and expect the users' mistakes and intercept them accordingly. In addition, the number of users quickly increases the desire for further functionalities.

Ultimately, it is a strategic decision whether one wants to go down this path and whether one can continue to do so in the future. This part of the simulation business is then pure software development and should also be carried out using the methods commonly used there.[1]

5.1.2 Commercial Product

The purchase of a commercial product will be the only possible way for most. Costs, support and compatibility with existing products are more important for the purchase decision than the underlying equation solution process. Very few will have sufficient experience in higher mechanics to be able to judge whether a formulation in absolute coordinates is now better or worse than that in relative coordinates. For the expert, these questions are serious, hidden in the program, invisible for the user, and he will be able to live with both.

Usually, only few tools are available to choose from for one's own questions. One certainty should be accepted from the very beginning. None of the products in question can cover 100% of the company's own requirements. The advertising and the professional product presentations will claim this (usually with small side blows on the unfortunately not so well developed competition products), but with a concrete employment one will again and again stretch one's limits. At this point, it is helpful to mandatory, if interfaces for own extensions and adaptations are available.

A benchmark can also serve as a basis for decision-making. The basic idea is that the potential simulation environments (or their providers) are confronted with a task from the planned use. After a reasonable period of time, the environment that solved the problem most easily, quickly or elegantly should be assessed. Unfortunately, due to the short time available or the limitations of the individual tools, it is often not possible to model a problem as complex as the task at hand requires. The relevance of the benchmark problem must therefore be questioned. A further risk is that benchmarks are deliberately or unconsciously interpreted in such a way that the tool preferred over the benchmark has an

---

[1]That is not meant as a depreciation, but as a message to let those deal with this part, who already know what they are doing. Almost anyone can quickly write a few lines of code, but not everyone is able to oversee the dimensions of such a project.

advantage. Despite all these efforts, this process can take several months to years. And despite this *objective* valuation of the benchmark, one should not be fooled by the fact that tool selection is often a management decision in which other criteria will also play a role.

However, since one binds oneself to such an environment for several years and a change to another provider is very time-consuming and cost-intensive,[2] one should take enough time to decide on the *right* tools. At relevant conferences, for example, it is possible to talk to users of the preferred products about their experiences with the products and their support in everyday life.

5.2   Basic Structure of a Simulation Environment

Basically, simulation environments consist of three essential parts. The model preparation (preprocessor), the actual calculation (solver) and the result preparation (postprocessor). The individual commercial tools are structured very differently in this respect. Either all three process steps are integrated in the tool or they consist of individual modules, some of which can even be replaced by products from other suppliers. The calculation of the system here refers to the establishment of the equations of motion and not the numerical approximation already discussed in Chap. 4.

5.2.1   Preprocessor

The so-called preprocessor is used to prepare the calculation. Both the creation of the model and its parameterization take place in this upstream process.

Most programs allow the graphical creation of models, i.e. individual elements such as joints, bodies, force elements, etc. can be created from prefabricated libraries (*multi purpose*) or complete modules, such as front axles, rear axles, drive trains, etc. (*single purpose*) and merge or configure them (Fig. 5.1). In this way, a basic model can be created quickly. The precision work of setting the details and the parameters can then be much more time-consuming.

A decisive difference for the handling of models, especially for experienced users, is whether the created model is stored by the program in binary form or as ASCII format in one or more files. Both methods have their advantages and disadvantages, which I would like to explain briefly.

5.2.1.1 Model Data in Binary Format
The storage of the model in binary form is the better way from a programming point of view. The files are usually much smaller and above all they are not editable by the user. The

---

[2]Sometimes, however, this also applies to the change to a newer version of the same tool.

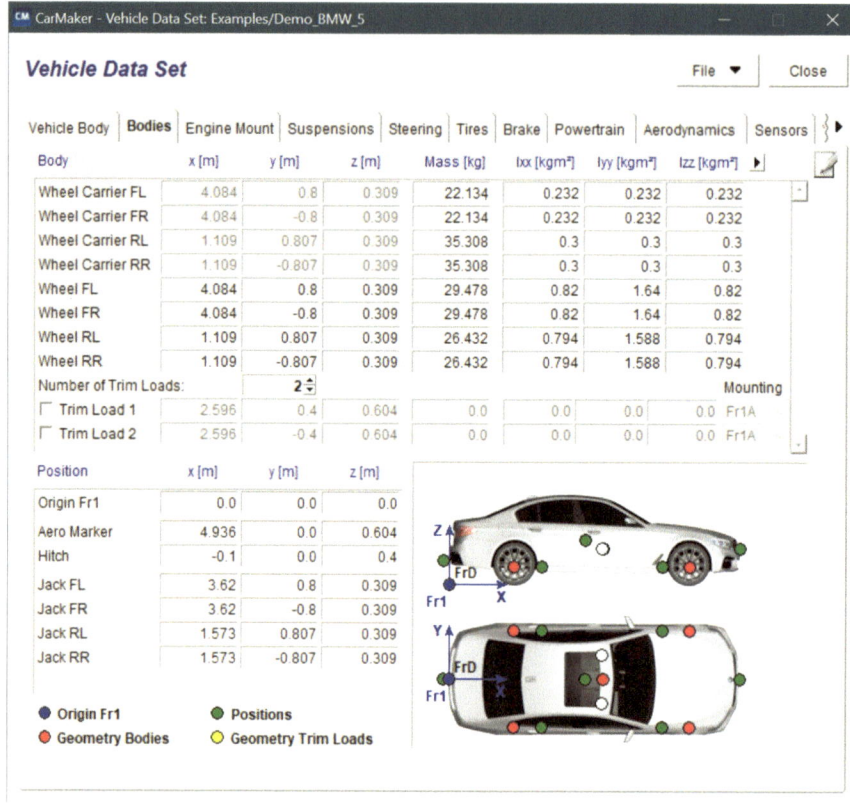

[IPG Automotive]

Fig. 5.1 User interface for vehicle configuration

software can therefore always assume that the data stored in the file is well-ordered and complete—the software technical verification of the validity of the data can be significantly leaner. For the user, this procedure means that changes to the model must always be made via the graphical user interface. Hopefully, this will also check whether the selected data format or the permitted value range has been adhered to. For the beginner this is usually better, because he is being guided and supported in necessary steps. Experienced users often find navigating through menu structures and input dialogs very annoying and time-consuming.

If different computer platforms or operating systems are used, it must be clarified in advance that the binary data can be used on all possible computers.

5.2.1.2 Model Data in ASCII Format
Alternatively, the model data can be stored in ASCII format, i.e. in a format that can be read and edited by the user outside the simulation program. The experienced users of such a

program appreciate this possibility, since they can implement modifications significantly faster at parameter or model level, depending on their experience, and get a larger overview of a part of the model. If you have to fight your way through several menu levels or dialog boxes to change a parameter or model setting, this way can be much faster.

Furthermore, this method is advantageous if you have to calculate many variants, as is the case with sensitivity analyses or optimizations, for example. You can then change parameters or model settings without calling the preprocessor and have the changed data calculated directly. Of course, many programs offer the possibility of specific variation of individual parameters, but often the necessary interfaces are not available, for example to connect an external optimization program of one's own choice. As far as the advantages of this approach to use the ASCII format.

A clear disadvantage is that most programs (rightly) expect a fixed format and a predefined sequence of the model data and only partially check compliance with these specifications. A single space character in the wrong place, a comma instead of a decimal point unintentionally inserted by the user, can cause a qualified error message in the best case, a crash of the program in the second case and a small unintentional modification in the model in the worst case. This case is therefore unfavorable because it may produce a calculation result that is not expected, but can still be explained. You may be chasing a vehicle phenomenon that is actually a model or parameter error. This method is therefore only recommended if you work with the greatest care and knowledge of the required format.

For cross-platform calculations (computer hardware/operating system), special characters (especially umlauts in German) or line ends can be interpreted differently, so that it must also be clarified in advance on which platform editing is permitted and which characters may be used.

5.2.2   Solver

The user of a simulation software does not have to deal with the equations of motion—unless he programs his own program and operates on the equation level. Usually, he sees a graphical representation of his model and in the background of the software it becomes a system of Ordinary Differential Equations (ODE) or Differential Algebraic Equations (DAE). Here then all equations of motion and binding equations are considered.

In the basic lectures of technical mechanics, one usually learns the equations of motion with the help of NEWTON's laws. Here, for each body within the system, there are six degrees of freedom and three external forces as well as three external moments, which are additionally limited by binding equations. An extension is then provided by LAGRANGE's equations of first or second kind, in which the bonds are formed by additional LAGRANGE's multipliers. Within the framework of higher technical mechanics, it is then possible to also use the procedures according to HAMILTON or APPELL or the use of quaternions. A good overview of these procedures can be found in [ScHB10].

Once the system of equations has been established, it is solved numerically as described in Chap. 4.

5.2.3   Postprocessor

After the calculations have been carried out successfully, you want to make the results visible. In this section, we will highlight the differences in whether the display is in the time or frequency domain, whether standard sizes or new sizes are to be displayed, and how important the animation tool is.

5.2.3.1 Representation in the Time Domain

As with a measurement of a real component or a real vehicle, signals (forces, accelerations, etc.) are also recorded in the model and displayed over time. A clear advantage of simulation is that, usually, a much larger number of signals can be accessed in the model than in a real measurement. Some simulation environments allow the amount of signals to be varied, with the effect that the result files only contain the signals that are interesting for the respective load case and thus the file size remains within limits. A disadvantage of this method is of course, as with measurement, that you can only access the recorded signals afterwards and cannot answer other questions that may arise later. This only leaves a new calculation with an extended set of output signals. However, if you always record all possible signals, the file size increases immensely and you drag around a lot of *unnecessary* information. In this case, a reasonable number of signals will be introduced over time, depending on one's own approach.

For complex problems, you will usually use multi-step procedures with step size or order control as integration procedures, which were already introduced in the last chapter. The step size control adapts the step size of the integration procedure to the respective situation. To put it simply: If everything is in a steady state, the step size is increased, if there is a jump excitation, the step size is reduced. What does this have to do with the representation of the simulation results in the time domain?

As can be seen in Fig. 5.2, the time points for the output of the simulation data generally do not lie on the time points calculated by the integration procedure. At this point, interpolation must therefore be performed between two integration points. This can lead to a situation in which the simulation results of the same model with the same integration settings but different output step sizes differ in detail. This means that different values may occur at the same output points, since the interpolation points may have been different depending on the interpolation method used. Mind you, this has no influence on the system behavior, only on the view we have on the system behavior. If you always use the same output step size, you will never encounter this phenomenon. However, I would advise you to try out the described effect to know how sensitive the combination of model, load case, integration method and simulation program reacts to this change.

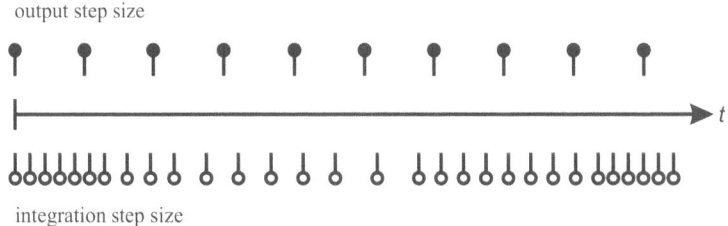

Fig. 5.2 Integration step size and output step size

A malicious function, especially for inexperienced users, can be the otherwise very comfortable auto scaling, i.e. the displayed results are automatically displayed in diagrams in such a way that they are adapted *ideal* into the diagram. Since the eye follows the curve first, drastic deviations may be visible here, especially when comparing two or more results. This can have different causes. First of all, the results may indeed show large differences. However, it can also be that the representation shows very small differences, which, if one goes back a step mentally, could also be interpreted as zero or as constant. Often, the view does not go at all to the ordinate and to the represented maximum value. Do two almost constant forces differ by 10^{-8} N in the time domain, this can look drastic with auto scaling, but without having any real influence on the result (Fig. 5.3).

5.2.3.2 Representation in the Frequency Domain

If vibration phenomena on the vehicle are considered, this will usually take place in the frequency domain. In the time domain, the differences between two variants are often not clearly discernible, if you look at the *wriggling* of two acceleration signals. A popular method is the *Fast Fourier Transformation* (FFT) which is used to convert the time signal into the frequency domain, for example to determine the power spectral density (PSD). Put simply, this is an attempt to replace a stochastic signal, for example, by the superposition of harmonic signals of different frequencies. The representation over the frequency then represents the weighting of the individual frequencies in the examined signal (Fig. 5.4).

Since humans perceive vibrations differently depending on their frequency, amplitude and direction of action, a variant in the frequency domain can be evaluated with the corresponding experience, for example with regard to ride comfort.

When comparing signals in the frequency domain, it is imperative that they have been transformed using the same procedure. What type of windowing was used? How wide is the window? Do the individual windows overlap? With the variation of these parameters the results can be displayed very differently. If the signals and not the procedures are to be compared, then these questions must be clarified in advance. A vivid and MATLAB related approach to this topic explains for example [Hoff11].

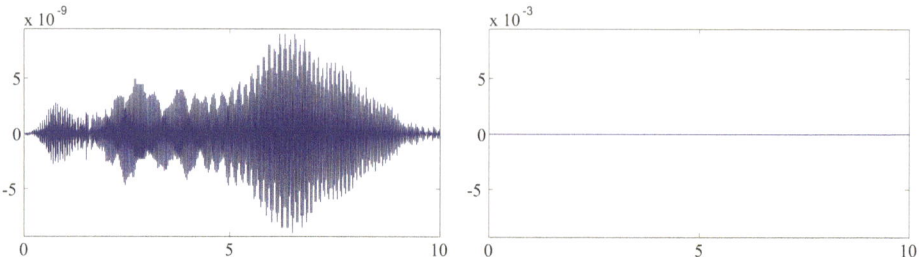

Fig. 5.3 Disadvantage of the auto scaling function

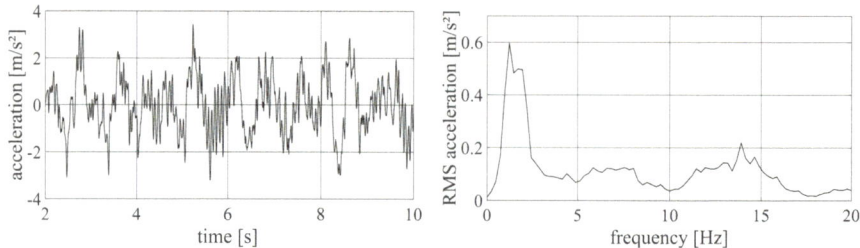

Fig. 5.4 Signal section in the time domain (left) and in the frequency domain (right)

5.2.3.3 Use of Own Measured Quantities

Sensors are necessary in the real vehicle to provide measured quantities for active systems or for measurement technology during the development process. If there is no sensor, the corresponding size cannot be determined either. The advantage of simulation models is that they provide a large set of measured quantities, in particular of forces, moments, paths, velocities and accelerations. If special values are required that are not included in the standard scope, they can usually be easily created.

As an example, the side slip angle shall be calculated, which can only be determined indirectly in reality. In the simulation model, the components in the individual spatial directions are usually available for all sizes, so that the side slip angle β simply from the following relationship of the components of vehicle speed in x- and y-direction can be calculated:

$$\beta = \tan\left(\frac{v_y}{v_x}\right) \tag{5.1}$$

Basically, there are two methods for the calculation of new quantities:

1. The quantity is calculated online, i.e. during the running simulation.
2. The quantity is calculated offline, only during the post processing.

In the first case, it is helpful if the simulation environment provides a mathematical calculation of existing signals, for example using a macro language. If this is not the case, a small program must be written for the calculation, which is linked to the simulation environment via a user interface.[3] This method is mandatory if, for example, a control algorithm or another vehicle component is to access this variable during the simulation.

If the simulation environment does not provide a simple possibility for the calculation of own quantities, the second possibility is often its calculation in post-processing. The recorded measuring channels can be offset against each other or filtered. The disadvantage is that these calculated quantities are not available during the simulation. In return, they can be post-treated and charged at will afterwards.

What to consider when supplying an active system with the sensor is discussed in more detail in Sect. 17.2. Here we limit ourselves to the evaluation of this measured quantity. Their presentation was dealt with in Sect. 5.2.3.1.

5.2.3.4 Animation

If the simulation software offers an animation of the results, a first impression of the time behavior of the system can usually be gained quickly and easily. If, for example, the calculation is cancelled halfway through, the animation can often be used to check whether the expected time behavior is present. Does the vehicle really drive on the given course or does it turn directly? If it literally flies apart because huge forces act at one point in the model, the cause is usually not physics, but a modeling error. For a more detailed analysis, the individual signals have to be dealt with and observed and checked for plausibility with the help of time series.

A detailed analysis with the help of animation is usually not possible, as the time series of individual signals or their evaluation in the frequency domain are more suitable. However, if you do not want to know whether different bearing forces occur between two variants, but rather which variant performs better in a driving maneuver, the animation can be very helpful. Especially if it offers the opportunity to see both results at the same time, such as the program CarMaker of the company IPG (Fig. 5.5). If this feature is used skillfully, it can also convincingly demonstrate the advantages of a variant to third parties who are not so involved in detail.

This aspect should not be neglected, because with the help of a suitable animation, the simulation results can usually be sold much better than with a representation in diagrams. There are often details to be seen that can usually only be interpreted by a very small group of experts. However, for you to be confident in the results, it is important that the basis of the animation is also the physics calculated by the simulation program. You can also use the animations to show a desired behavior, but then you get close to the animation artist and that is not the subject of this book.

---

[3]Often described as the so called *user code* or as *user defined function.*

[IPG Automotive]

Fig. 5.5 Comparative animation of two simulation results

Complex representations with rendered surfaces and a detailed environment are especially impressive for non-specialists. But they can also distract from the actually important points.

5.2.3.5 Further Evaluation Procedures

The group of multi purpose tools often offers the possibility of using additional evaluation methods. This includes, for example, kinematic analysis, in which forces and moments are not taken into account and a pure motion analysis is performed. This process is particularly popular in the design of suspensions.

Another possibility is linearization, which can be used to determine the eigenvalues and eigenvectors of a system. The experienced computational engineer can already see whether the values are in the expected range. Parameterization errors of mass and stiffness properties can thus be detected.

The last procedure to be mentioned is static analysis, which can be used to determine the static loads, for example to determine preloads or to adjust the vehicle level.

Primarily, however, it is a matter of dynamic analysis in which the motion behavior and the effect of forces and moments have priority.

5.3   Interfaces for Co-simulation

Today, many simulation environments already offer interfaces to other simulation programs in their standard scope—some, however, only for an additional charge. Often an interface to MATLAB/SIMULINK can be found so that control engineering issues from the *mechanical* model can be outsourced to the block-oriented environment. Depending on the environment, up to three types of co-simulation are offered, which are briefly introduced in the following sections.

5.3.1   Controller Import

If you have the corresponding toolboxes (e.g. SIMULINK CODER), you can export the controller as C code and then integrate it into the MBS environment (Fig. 5.6). The advantage of this method lies in the fact that you only have to deal with one simulation environment. The MBS computational engineer remains in its familiar environment and is supplied with the controller by an expert (or supplier). This can also be in the form of a binary library to protect the contents of the controller.[4] The control of the simulation process then lies in the MBS environment. The controller code is integrated in the normal program sequence. The total computing time will change only slightly.[5] The simulation process usual for the MBS computational engineer runs as known. During runtime he only needs one license for his MBS environment. The license for the controller development environment was used only during the controller design or during its export.

This is particularly useful if the controller is fully developed or if it is to be changed at longer intervals and is to be used together with the MBS model. If the controller is still being further developed, the export from the controller environment and the import into the MBS environment must be repeated again and again. Depending on the working environment and the distance between the MBS computational engineer and the controller developer, this can lead to delays in the development process.

5.3.2   Importing MBS Models

Some simulation environments can export the entire MBS model as source code, which can then be exported as S-function in SIMULINK (Fig. 5.7). However, this variant is much less common. The opposite of the above applies here. This variant should be used (subject to availability) if the system, i.e. the MBS model, does not need to be changed and the controller is to be further developed. The control of the simulation process then lies in the controller design environment. The above applies to the calculation time. The typical simulation process for the controller developer runs as usual. During runtime he only needs one license for his controller design environment. The MBS license was only used during the model creation of the MBS model and during the export.

---

[4]Depending on the operating system, i.e. as DLL or LIB.

[5]Based on the calculation time for the controller algorithm and the necessary communication. The changes of the entire system behavior can of course lead to an increase in calculation time.

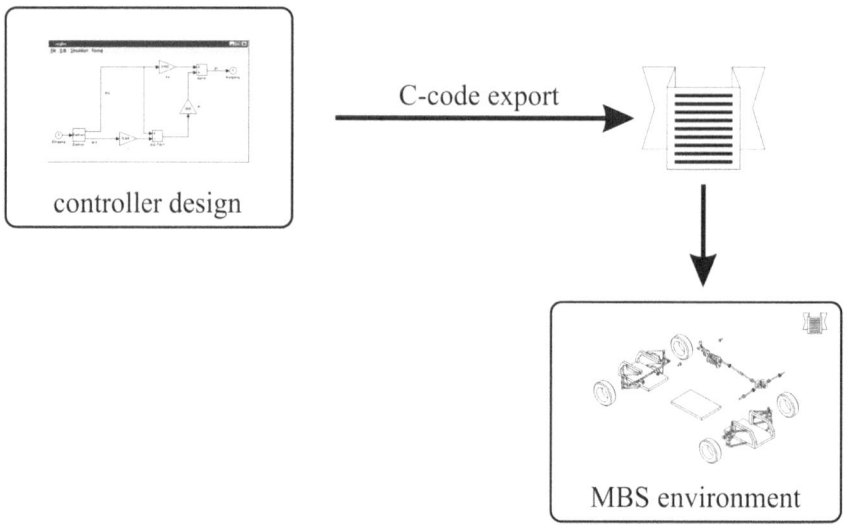

Fig. 5.6 Controller import into a MBS environment

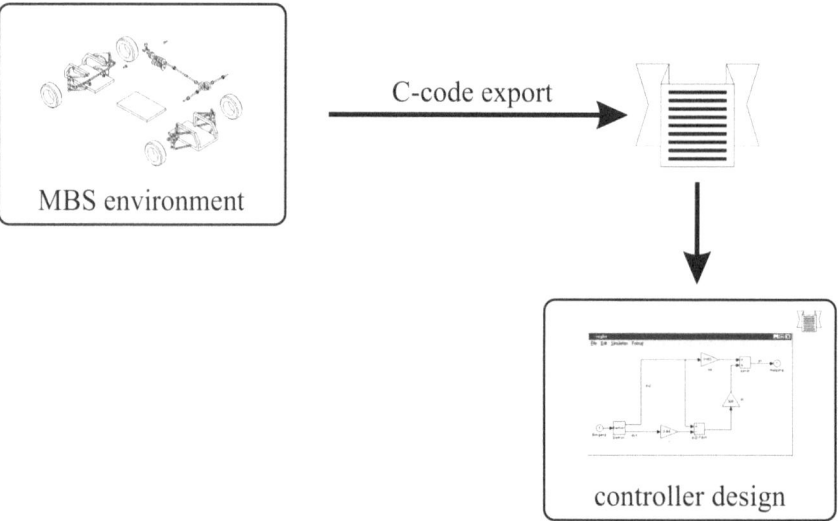

Fig. 5.7 MBS model import into a controller design environment

In these scenarios I assume that the developer of the MBS model and the developer of the controller are not one and the same person. Should this be the case, the procedure will somewhat be simplified.

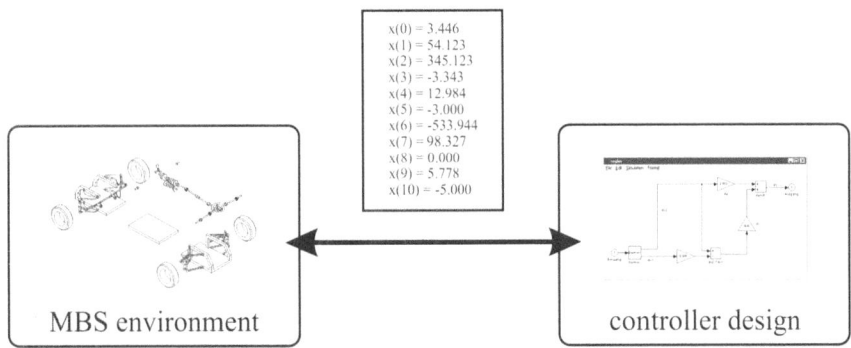

Fig. 5.8 Online simulation

5.3.3   Online Simulation

The most complex of the three possibilities is the online co-simulation. The MBS model runs in the MBS environment and the controller in the controller design environment (Fig. 5.8). At defined communication times, both programs exchange a state vector, in which, for example, sensor signals for the controller and control signals for actuators can be contained in the MBS model. SIMULINK also provides S-functions for this purpose.

The user of this option should be able to handle both simulation tools and also needs licenses for both so that the programs can run simultaneously. The computing effort, to which a communication effort may be added under certain circumstances (see next section), is highest with this variant.[6]

5.3.4   Potential Communication Problems

Regardless of the type of coupling, you have to think about time management and data exchange at the beginning. At what times are data exchanged and what data formats does the other side expect? Are larger steps in the value range permitted or must a continuous course of the signal be guaranteed?

For all variants, a state vector must be exchanged between the controller and the controlled system in which information is exchanged between the model parts concerned. Experience shows that the dimension and the configuration of this vector (at which position is which state variable) is changed in the course of development. A quick comparison with

---

[6]Of course there may be cases in which both model parts are so complex that an allocation to two processors offers benefits in calculation time despite the communication effort. These, however, appear to be the exception to me. This does not describe a *real* parallelization of the calculation process.

an older release is then no longer possible without further ado. Here it is recommended to introduce a version number of the releases and to exchange it between the model parts. In this way, functions that require additional variables can be switched off if necessary. Alternatively, the evaluations of the state vector can be designed differently depending on the version.

As an example for the time behavior we take a hydraulic actuator, which manipulates the suspension spring in its spring travel, as it is the case with the *Active Body Control* at some Mercedes cars.

Let us imagine a solution in which the controller and the actuator have been modeled in SIMULINK and the rest of the chassis in an MBS environment (Fig. 5.9). If the system dynamics of the hydraulic actuator have been neglected as a first approximation or only rudimentarily represented, it can carry out its deflections very quickly. The body spring in the MBS model does not care, since it only produces a displacement-dependent force. In the same force path, however, there is also the damper, which reacts speed-dependently with its force effect. If there are now major differences in path between two communication points, this inevitably results in a very high speed. The damper would generate unrealistic forces—the overall system would not function as expected. If the signals are exchanged in millisecond intervals in order to be supposedly as accurate as possible, a displacement change of 1 mm per time cycle already results in a speed of 1 m/s, which already feels like a pothole to a damper.

It is therefore necessary to model the system dynamics correctly in order to steer the rate of change into realistic paths. This is a trivial statement in itself, but it can become a trap for beginners if they concentrate too much on the controller design and too little on the mechanics.

A further question is how often the data must be exchanged between the two model parts and in what rhythm. The more often this happens, the greater the communication effort. If both simulation programs run on a processor with several cores, the time for the exchange is short. If the two simulation programs are accommodated on two different computers that communicate via a network or a data bus, the time required increases drastically.

The exchange rate depends on who is leading the process. Let us first assume that an integration procedure with step size control is used. As a consequence, one does not know in advance how large the step size will be in each case. Would you carry out a transmission in every or every n-th time step, the distances between two communication points would not be equidistant. However, controllers usually communicate with a fixed clock rate (Fig. 5.10). A time management system is therefore required that exchanges data between the two parts at fixed times. Irrespective of whether this point in time lies at the end of an integration step or in the middle of it. This means that interpolation or extrapolation may be necessary. Simply using the last valid integration step, no matter how long it has been available, can lead to discontinuities in larger distances and highly dynamic systems (e.g. hydraulics). This in turn can lead to unwanted system behavior.

Fig. 5.9 Interface between controller and chassis

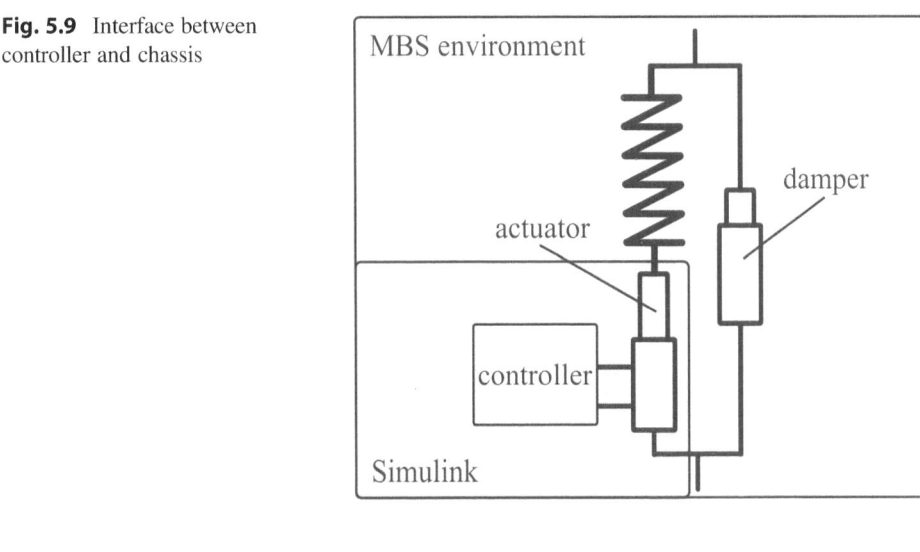

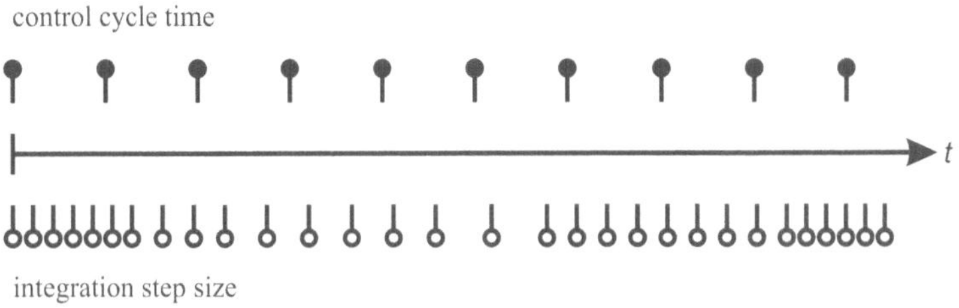

control cycle time

t

integration step size

Fig. 5.10 Integration step size and controller cycle time

One could now object that there are also integration procedures with fixed step sizes. However, whether these then do justice to the mechanical problem must be examined more closely in each individual case. Information on this can be found in Chap. 4.

Most commercial programs do not force you to deal with this issue, but they often do not tell you how they solve it. Hopefully, these examples show that it is worthwhile to start with the signals that are exchanged and their transmission.

References

[Hoff11]   HOFFMANN, J.: *Spektrale Analyse mit MATLAB und Simulink. Anwendungsorientierte Computer-Experimente*. Berlin, Oldenbourg Wissenschaftsverlag, 2011

[ScHB10]   SCHRAMM, D., HILLER, M. AND BARDINI, R.: *Modellbildung und Simulation der Dynamik von Kraftfahrzeugen*, Springer, Berlin, 2010

Simulation Process

6

Since that which is compounded out of something so that the whole is one, not like a heap but like a syllable—now the syllable is not its elements, ba is not the same as b and a, nor is flesh fire and earth.

Aristotle, Metaphysics

If the simulation is a tool of an individual who is looking for a supporting tool for his development activity, he may not communicate the results of the calculations at all, but use them as the basis for his decisions. If you work this way, you can skip this chapter. However, you should remember its existence when the need to simulate becomes ever greater and the results become the basis for the decision of third parties. Then the procedure described above is no longer sufficient.

When working in teams with the same models or when the results of calculations are requested and used by others, a more structured approach is advisable. There is not one procedure for all, but it must be adapted to the respective company and work structure. If the simulation results are an integral part of the development process and are clocked into the schedule, this has a different quality than if there is a need for simulations in loose succession for individual questions.

The cost advantage over the test is often mentioned as the advantage of the simulation. This is an economically comprehensible argument, nevertheless it is usually wrong. If the simulation is integrated into the development process and reliable results are expected from it, its costs cannot be neglected. The prices for the necessary hardware have fallen sharply, but the license costs usually remain very high. Moreover, real simulation experts are rare and expensive. But all this is only a small part of the total costs. The much larger part—which is rarely determined—consists of the cost of obtaining the necessary parameters for

© Springer Fachmedien Wiesbaden GmbH, part of Springer Nature 2021
D. Adamski, *Simulation in Chassis Technology*,
https://doi.org/10.1007/978-3-658-30678-6_6

the models and this in an often high quality. Often, the usual measurements of components, as required by component developers, are not sufficient for sophisticated modeling. This means additional or more complex measurements. In addition, there is a data management process that ensures that the *right* and current data are available for the simulations. At the beginning of the century, OEMs made great efforts to establish these processes in their companies. Cost driver is the simulation principle, which is formally called GIGO (*Garbage in—Garbage out*) and, less formal, *Shit in—Shit out*. The best vehicle model is useless if the input data are of inferior quality. The less I know about the component to be simulated or even the vehicle to be simulated, the more uncertain are the results. This should not be understood as an argument against the simulation, but rather as a plea for a careful preparation of the necessary data. This follows as a modeling principle:

Only good input data provide good output data!

As will be suggested in Sect. 6.5 on reproducibility, it is immensely difficult to recreate a result that may have been produced a year ago. This is directly connected to the significance and productivity of simulation in a company. OEMs have developed a lot in this sector in recent years, as you can read in [BaGK08], [BrDG06] or [Pohl10].

6.1   Parameter Procurement

It is an essential difference whether a parameter is already fixed, for example because it necessarily results from a specification or geometry (e.g. the wheelbase) or whether it is a design parameter that is only to be determined by the simulation. This distinction leads to the requirement that all necessary fixed parameters must be available before the first calculation.

6.1.1   Need of Parameters

You need parameters for each model. If the previously used single-mass oscillator requires only three parameters (mass, stiffness and damping), complex complete vehicle models can require four-digit parameter numbers. The question is, where do these parameters come from? Only a few chassis-relevant parameters are known in the early concept phase. For a vehicle these may be the wheelbase, the target weight or the engine power (and thus the requirements from the maximum speed), for a component the available installation space may already be fixed. If initial calculations are to be made at this stage, most of the data will be based on estimates or experience with a possible previous version.

The clearer the objective of the product becomes, the more data can be derived from the construction or there may even be first models which can already be measured. At least the relevant geometry data can be read from the CAD system. If the constructions are provided with the correct material data (density), the mass data (mass, moments of inertia, center of gravity) can also be taken. There is a more detailed description on this subject in Sect. 6.1.6.

If you are not responsible for the design or measurement, you need a data supplier. Either in-house or externally. Here one will have to specify exactly in which form one needs the data. Up to what force level should elastomer bearings or up to what speed should dampers be measured? In what form should the data be delivered? If you receive a fax (also scanned by email),[1] you have to transfer all measuring points by hand into the electronic form. Usually, the data is electronically available to the measuring person. The measuring programs can also be exported to non-proprietary data formats.

When modeling a chassis component, it should be noted that the more detailed the description, the more parameters are usually required to calculate a specific component. The effort involved in measuring these parameters must always be kept in mind. This is not just a financial question. It may also prevent you from getting this data more than once. This is especially true if the measurement procedures are not standard. At least one can assume, however, that the procurement of the data takes longer than with a standard process.

The fewer parameters are required and the easier it is to obtain these parameters, the safer it is to completely parameterize the component. A common procedure, once a model has been fully parameterized, is to first provide successor models with the parameters of the previous model and only reassign those parameters that are also obtained from new measurements. All other parameters remain *old*. At first, this is a thoroughly tried and tested procedure, but it quickly leads to sets of data that can only be described as inconsistent. A commentary of the parameters, which also includes the parameter setting date and the source, is incessant. Unfortunately, not all programs offer this possibility and only few users are consistent enough to do it.

It is also a management decision that care takes precedence over speed here. This is an investment in the future that has already been *forgotten* in day-to-day business by both the computational engineer and the management.

6.1.2   Naming of Parameters

Once the separation of model and parameters described in Sect. 3.9.2 has been completed, it is necessary to consider how to identify these parameters—the so-called nomenclature. Before starting, it is worth taking the time to understand how many people of different languages are supposed to work with the model and how many are supposed to parameterize it. Here, too, the same applies—the effort can be minimized for one-time use. However, as soon as the model or model family is to be reused and used in many variants, a uniform naming of the parameters, from which everyone can immediately read out the meaning of the value, is mandatory. Many misunderstandings can be avoided and troubleshooting is drastically simplified.

---

[1]Older people know what I mean.

The first principle is that a parameter name must appear only once in the model. This seems natural and very few will doubt that it is necessary. However, this principle can be broken very quickly if several sources are used to create an overall model. If the developers of the sub models have not agreed, such overlaps may occur. If the software used does not check the multiple occurrence of a parameter name, the value used depends on the reading routine of the software. An obvious mistake that is not easy to find. There is no *one and only* approach that represents the optimum for all applications. It is more important to reach a consensus on the use of a uniform naming convention. I will show a few examples from which you can derive your own naming strategy.

Now, what about is meant by *uniform naming*? First of all, the location of the parameter, i.e. the associated component, should be indicated by its name. On the one hand there is the possibility to use so-called descriptive names, which means immediately readable and interpretable designations.

Example 1: DÄMPFER_VORNE_LINKS or shorter DÄMPFER_VL

With this example you can catch several problems.[2] This naming assumes that only German-speaking persons have to work with this parameter. Furthermore, it assumes that the programs used can handle umlauts on all computer platforms (operating systems). If this is still the case for your own company, it can quickly reach its limits with suppliers or service providers who may be involved. The underscores instead of spaces are already a concession to some operating systems or programs. Let us first assume that the German language is sufficient and at least defuse the umlaut question on the following proposal:

Example 2: DAEMPFER_VORNE_LINKS or Daempfer_Vorne_Links or Daempfer_vorne_links or daempfer_vorne_links or shorter *_VL

It must also be clarified whether the software to be used is case-sensitive. This is handled very differently. In order to avoid this problem, the way in which one writes in capital letters and when one writes in small letters must be determined. Both can be used. For example, you could use completely uppercase parameters for constants and completely lowercase parameters for calculated parameters. One would immediately recognize their type because of their spelling.[3] But since we are still at the location and not at the type of the parameter, we now agree on the following suggestion:

---

[2]I decided to leave the original German example (DAMPER_FRONT_LEFT) in the English text, because everyone whose mother language is not English will find common examples in his own language with the same problems. The others may have colleagues in other countries.

[3]One calculated parameter i.e. is the current spring length, which can be calculated from the two binding points (which should be constant as parameters).

Example 3: `daempfer_vorne_links`

The clear advantage of knowing immediately where the corresponding parameter belongs can also quickly develop into a tapeworm if you look at the following example:

Example 4: `upper_control_arm_bearing_front_front_left`

If the physical type is added, as recommended below, the parameter can be quickly called

`upper_control_arm_bearing_front_front_left_radial_stiffness`

It is questionable whether this will then contribute to readability.

Alternatively, you can use short names that require a translation table so that the assignment is unique. So the upper front control arm bearing on the left side of the vehicle could be turned into a `uca_f_fl`.[4]

In both cases, it is important that the order of the naming is determined. In this case: component—local component location—global component location.

In addition to the location where the parameter is used, the physical type should also be identifiable. Computer science can serve as a role model here. The classic example is the so-called *Hungarian Notation,* where a variable is preceded by its type as an abbreviation (prefix). Based on this procedure, the physical type of the parameter can be coded. This allows a stiffness to be achieved with the usual mechanical c, Cartesian position parameters with x, y and z can be added. Here the decision is to be made whether the abbreviation should be placed in front or behind. If the physical type is prefixed as an abbreviation, as in the case of the *Hungarian Notation* you sort the parameters alphabetically, you can, for example, see all stiffness at once (Table 6.1).

The clear disadvantage is that one cannot get all component parameters listed together by a simple sorting algorithm. For this it would be useful to set the physical type to the end of the parameter. A simple sorting then brings all component parameters together—with the problem of not getting an overview of all stiffness (Table 6.2).

It is also possible to integrate the unit to be used into the name—this could be used if there is no comment option. But here it seems more sensible to me to make a clear agreement to use all data in SI units, more on this in the next section. All these agreements on naming conventions and units should be set down in writing accessible to each user. Especially when abbreviations are used instead of descriptive names and third parties, especially external parties, are to work with them.

---

[4]Nearly all simulation environments mirror the left wheel suspension side onto the vehicle's right side, so that one could forgo the addition vl. If one wants to keep the freedom of building the vehicle unsymmetrical or using the measured (different) values of the left and right bushing, one should equip both sides with their own parameters. Even if they will be identical in 99% of cases.

Table 6.1 Parameter list with prefix[a]

Name	Value	Comment
c_ax_lca_f_fl	28,000	N/m, status Q2, 06/21/13
c_ax_lca_r_fl	27,000	N/m, status Q2, 06/21/19
c_ax_tr_fl	20,000	N/m, status Q3, 06/28/13
c_ax_uca_f_fl	25,000	N/m, status Q3, 06/28/19
c_ax_uca_r_fl	20,000	N/m, status Q3, 06/28/19

[a]The depiction is exemplary and varies from program to program

Table 6.2 Parameter list with suffix

Name	Value	Comment
uca_r_fl_c_ax	20,000	N/m, status Q3, 06/28/19
uca_r_fl_c_rad	17,000	N/m, status Q3, 06/28/19
uca_r_fl_x	120.1	mm, status Q2, 06/21/19
uca_r_fl_y	−687.3	mm, status Q2, 06/0619
uca_r_rl_z	412.2	mm, status Q3, 06/28/19

6.1.3   Unit-Related Parameters

In the majority of cases, parameters reflect physical quantities, which in most cases are assigned units. Large sources of error are missing or incorrectly converted units. Unfortunately, not all programs offer sufficient support for units here either. Those which do, will automatically convert to the default units, no matter which unit the user specifies. Where there is no support at all, standards (e.g. SI units) must be strictly adhered to, but these must also be clearly defined and made known. Unfortunately, the direct transfer of measured values also offers a major source of error, since very different units can be used depending on their origin (sometimes depending on the age or supplier of the measurement software or hardware). If the simulation software provides a commenting of parameters, it is advisable to mention the used unit in the comment and to keep the comment up to date in case of changes.

6.1.4   One-Dimensional Parameters

Many parameters can be represented by constants. Parameters such as weight, position or length can be represented as a scalar, i.e. one-dimensionally. However, there are also sizes that can occur in one or more dimensions, depending on the application. The position is physically a vector or the mass inertia tensor is a matrix. But few programs offer the possibility to use these data formats, so that the individual components of the vector or the matrix are stored in their own scalar parameter.

For example, if the component has a linear force behavior (Fig. 6.1), which means the stiffness is not dependent on the path, it is usually sufficient to represent this stiffness by a scalar parameter.

Fig. 6.1 Linear spring stiffness

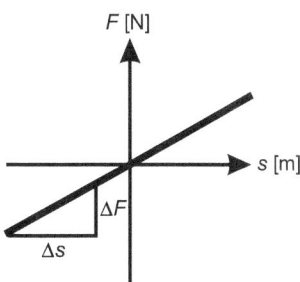

$$c = \frac{\Delta F}{\Delta s} \tag{6.1}$$

However, it should be noted that the component is only deflected in this linear range within the simulation. If this is not the case or cannot be predicted with certainty, stiffness changes (e.g. progressions) may be neglected. They can then lead to very large and usually impermissible deflections or force levels.

If the vehicle behavior is to be evaluated in the frequency domain, all variables must be linearized. A procedure for a quarter vehicle model, for example, is described in [Rösk12].

6.1.5 Multidimensional Parameters

If the physical quantities can no longer be described by a value, they must be represented either by a functional context, by characteristic curves or maps.

6.1.5.1 Characteristic Curves

If, for example, the stiffness of a component changes over the path, these changes must be described by a characteristic curve (Fig. 6.2). The most common are force-stroke curves, since the force laws for stiffness are often defined in such a way that the input is the stroke and the output is the force. The stiffness is then the derivative of this curve over the path.

It would also be possible to display the stiffness curve over the path. However, the force would then have to be determined by integration. Which is rather unusual. After all, this procedure can be very helpful in preparing the characteristic curve. If, for example, the wheel-related stiffness of the unit consisting of a spring, bump stop and a tension stop spring is considered, then the stiffness changes at certain points of application during compression and deflection are frequently defined here.[5]

By looking at the stiffness curve over the spring stroke, these points of application are much easier to identify. If a wheel-related measurement of the force-stroke curve is

---

[5]More detailed information to be found in Sect. 9.3.3.

Fig. 6.2 Non-linear spring
stiffness

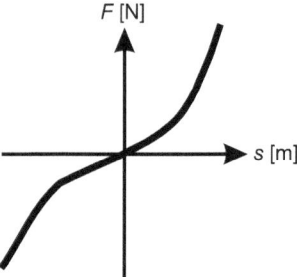

obtained, it is possible to identify the points of application by simple differentiation.[6] In the force-stroke curve, this is often much more difficult.

Figure 6.3 shows a wheel-related characteristic curve in which the tension stop spring starts at a deflection of 30 mm. The bump stop starts at a spring deflection of 20 mm.

If values are required between the measuring points or outside the measuring range, they must be interpolated or extrapolated. Numerical methods for interpolation or extrapolation are described in Sect. 4.5.

6.1.5.2 Characteristic Maps

If a characteristic curve is the two-dimensional representation of a function with one parameter, the extension by at least one further parameter is referred to as a characteristic map. The organization of the data depends strongly on the program used. While the parameters of a characteristic curve are often stored as value tables (Table 6.3), this can be done in several ways for a characteristic map. In this example of a (shortened) damper characteristic curve, a unique force can be determined for each speed. Values between the sampling points must be interpolated, extrapolated outside the measuring range.

The parameters of characteristic maps can be stored in a matrix form or as a combination of several characteristic curves. The matrix form is suitable if all characteristic curves are used with the same interpolation points (Table 6.4). In this example, the force can be determined if values for the velocity v and for the current i are available. The interpolation or extrapolation must now be two-dimensional.

If characteristic curves are available from measurements taken at different sampling points (Table 6.5), there are several ways of proceeding with these data. If the program used requires the matrix form, the data must be converted once in advance to the same grid points. With programs like MATLAB you can do this relatively comfortably, but with a little preparation you are also able to get to the matrix form with every spreadsheet.

---

[6]Differentiating measured curves often leads to heavily noisy results. Those who want to filter here, needs to know exactly what they are doing, as most filters *smear* the signal and thus defer the insertion points. Alternatively, zooming in on the interesting value section can be helpful sometimes.

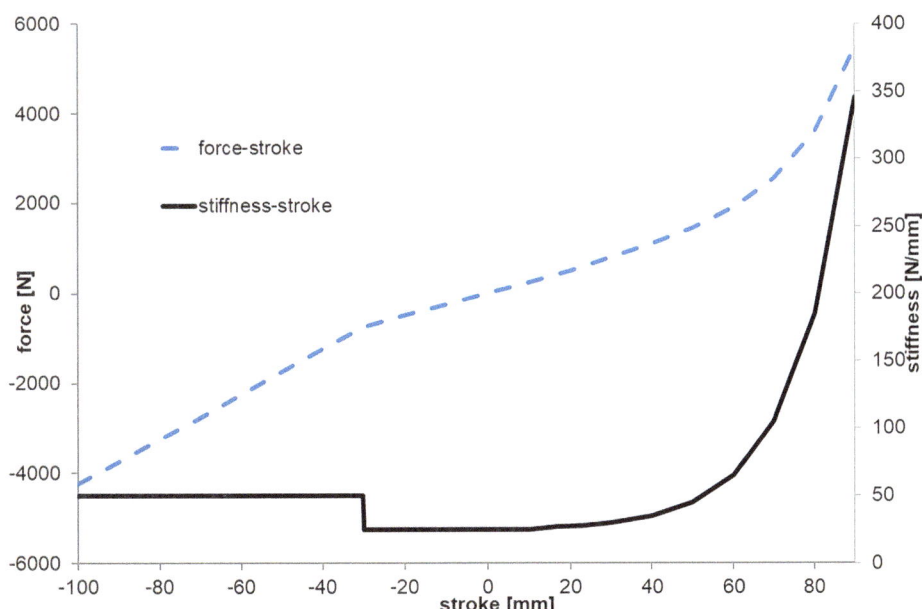

Fig. 6.3 Force-stroke and stiffness-stroke diagram

Table 6.3 Table of values

v [m/s]	F [N]
−0.5	−1000
−0.1	−400
0.0	0
0.1	1100
0.5	3120

Table 6.4 Characteristic map as matrix

	i [A]				
v [m/s]	0.2	0.4	0.6	0.8	1.0
−0.5	−600	−800	−1000	−1200	−1400
−0.1	−240	−320	−400	−480	−560
0.0	0	0	0	0	0
0.1	660	880	1100	1320	1540
0.5	1872	2496	3120	3744	4368

So, you have the effort of the conversion once and afterwards you can use the offered standard routines of the used program. A clear disadvantage, however, is that the measured original data and the model data used differ. At first glance, it is not possible to ensure that

Table 6.5 Characteristic map from individual characteristic curves

i = 0.2 A		i = 0.4 A		i = 0.6 A		i = 0.8 A		i = 1.0 A	
v [m/s]	F [N]	v [m/s]	F [N]	v [m/s]	F [N]	v [m/s]	F [N]	v [m/s]	F [N]
−0.42	−528	−0.61	−932	−0.51	−1000	−0.55	−1290	−0.48	−1358
−0.15	−360	−0.23	−736	−0.10	−400	−0.17	−816	−0.20	−1120
0.00	0	0.00	0	0.00	0	0.00	0	0.00	0
0.15	812	0.25	1486	0.13	1100	0.12	1441	0.11	1611
0.45	1685	0.53	2646	0.57	3120	0.56	4193	0.47	4106

the correct measurement is used in the model. A directive on the designation of data statuses (and their strict compliance) must rule out any confusion.[7]

The safer way therefore would be to use the measured data unchanged—but most programs do not allow this, since this directly affects the internal interpolation and extrapolation routines. If you have the possibility (and the ability) to implement your own routine here, the risk of errors could be reduced.

6.1.5.3 Number of Grid Points

Depending on the measuring procedure, the characteristic curve or the map may be very finely scanned (Fig. 6.4).

If this high-resolution characteristic were used in the simulation, this could lead to high memory requirements and also to a slowdown of the calculation, depending on the program. This would not improve the accuracy of the statement. For this reason, the characteristic curve used in the simulation should be as accurate as necessary and as coarse as possible. If the resolution is too low (Fig. 6.5), the problems with interpolation or extrapolation discussed in Sect. 4.5 may arise.

If the calculation is not performed on the own workstation, but on a central computing node, the copy process of the required files on the computing node is added in addition to the actual CPU time. If a large number of short calculations have to be carried out, as is often the case with optimizations, this proportion can increase considerably and contribute significantly to the overall duration of the simulation.

If the simulation software does not require an equidistant representation of the characteristic curve, the more linear components can be resolved very roughly and the components with large changes in the stiffness curve (progression) can be resolved somewhat more finely (Fig. 6.6). Otherwise, one has to orient oneself more to the nonlinear courses and accept a larger number of grid points.

---

[7]Using general life experience and with Murphy's Law in mind, errors that can arise, will arise. Therefore, a potential source for error lies at this point.

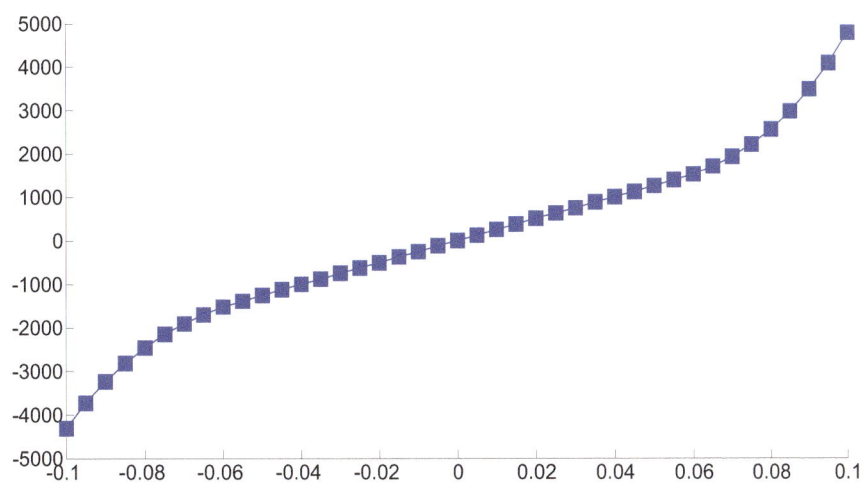

Fig. 6.4 Characteristic curve with high resolution

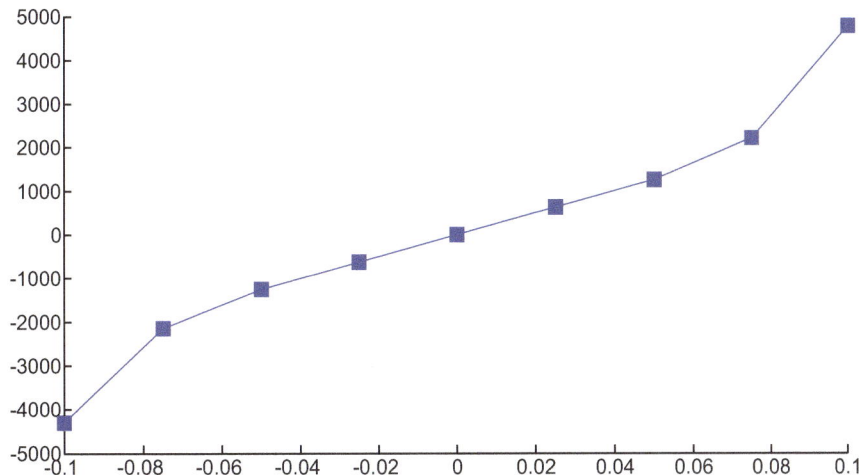

Fig. 6.5 Characteristic curve with low resolution

6.1.5.4 Parameters Affected by Hysteresis

If the measured characteristic is affected by hysteresis, this effect often cannot be transferred directly to the model, since it usually is due to friction or damping effects in the component and must therefore be described with a different force law (see Chap. 10). In order to filter out the force-stroke characteristic curve from this measurement, it is useful here to identify the mean characteristic curve which lies between the upper and lower branches of the hysteresis affected characteristic curve (Fig. 6.7). If this mean characteristic curve is then superimposed with a friction or damping effect, the hysteresis is well represented.

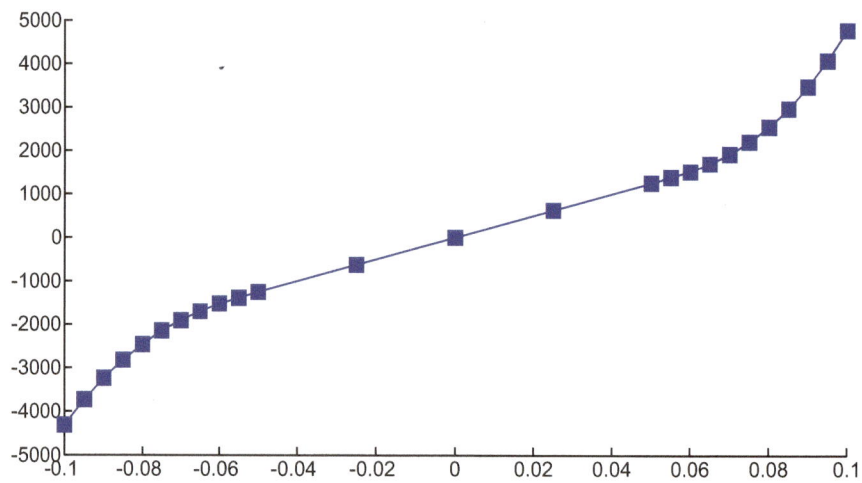

Fig. 6.6 Characteristic curve with demand-oriented resolution

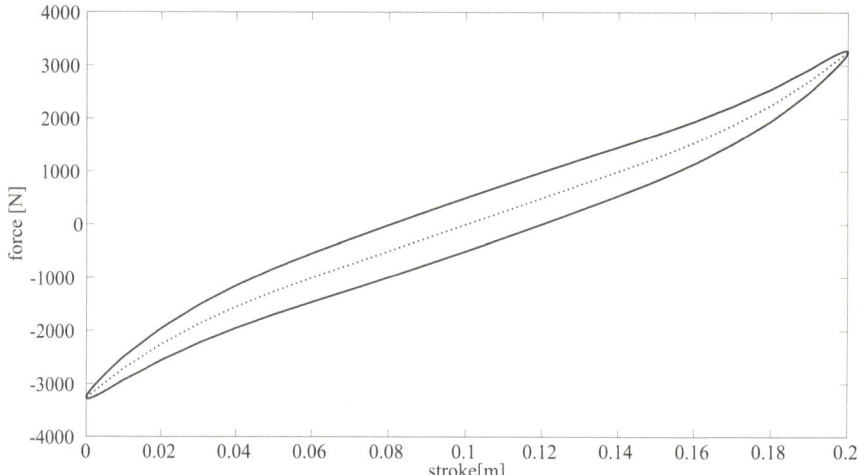

Fig. 6.7 Hysteresis affected characteristic curve

6.1.6 Vehicle Reference System

An essential question to be clarified before using simulation software for the first time is that of the used vehicle reference system.[8] Can you define it yourself or is it predetermined by the program? Where is it located and in which direction is it defined? Cartesian

---

[8]By vehicle reference system the mobile, vehicle-fixed coordinate system is meant, which all vehicle coordinates correlate to.

Fig. 6.8 Different vehicle
reference systems

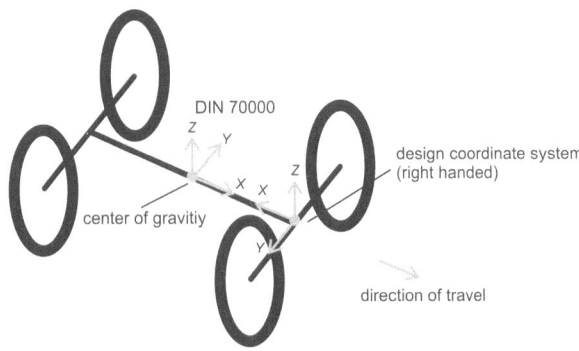

coordinate systems are usually used, and in the automotive industry the most common position of this reference system in design is the center of the front axle. In order to keep the majority of the coordinates positive, it is oriented against the direction of travel so that all points in the direction of travel (x-direction) are positive behind the front axle and negative in front of it. Since a considerable part of the parameters are obtained from design data, it is worthwhile to choose the same definition of the vehicle reference system in the vehicle model as in the design. That way, no conversion is necessary (Fig. 6.8).

However, there are other specifications from the respective programs, so the reference point can also be located behind the vehicle on the roadway level, oriented in the direction of travel. The conversion to this point is to be solved by the usual means of a spreadsheet, but then one uses *unnatural* coordinates that cannot be directly compared with construction parameters.

If an additional reference system is inserted for the rear axle, all coordinates of the rear axle can be referenced to the center of the rear axle, i.e. in which x-coordinate, the wheelbase of the vehicle does not appear. The rear axle system is shifted by the wheelbase from the front axle system. In this way, different wheelbases (extended variants) can easily be displayed without having to adjust the coordinates.

The definition according to [DIN70000] contradicts this procedure. Here the x-coordinate is oriented in the direction of travel and the center of gravity is usually indicated as the location. This is rather the view of the mechanics, because the centrifugal force during cornering or the weight force is placed there and then converted to the axes. Since the center of gravity depends on the motorization and load, it is rather unsuitable as a reference system.

For international projects, it is necessary to agree whether to use a system according to [DIN70000] or a system according to [SAEJ670e], as this has an influence on the signs of the y- and z-axis and the corresponding sense of direction of the rotations about these axes (Fig. 6.9).

Fig. 6.9 Coordinate system
according to DIN or SAE

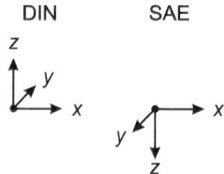

6.1.7   Mass Properties

Is the coordinate system in the CAD system identical to that in the simulation program or
do parameters have to be converted? This applies in particular to mass moments of inertia,
which, as we have learned in technical mechanics, are related to an axis of inertia. In the
calculation programs, these often go, through the center of gravity of the body. In the CAD
world, there is a component origin that is often used as a reference system (Fig. 6.10). A
conversion is, thanks to the Swiss mathematician JACOB STEINER, not difficult for homogeneous mass distributions (for example the Eqs. (6.2)–(6.4) for the main moments of
inertia)—but it must be performed.[9]

$$\Theta_x^S = \Theta_x^0 - m \cdot a^2 \tag{6.2}$$

$$\Theta_y^S = \Theta_y^0 - m \cdot b^2 \tag{6.3}$$

$$\Theta_z^S = \Theta_z^0 - m \cdot c^2 \tag{6.4}$$

Often, there is no homogeneous mass distribution because the component consists of a
variety of different materials. The data contained in the CAD system may be incomplete
here if these different materials have not been assigned. If the influence of the mass moment
of inertia is classified as significant, the component must be oscillated. If the simulation
environments do not test this anyway, there is a simple possibility to test the plausibility of
the values of the inertia tensor (Eq. 6.5).

$$\Theta^S = \begin{bmatrix} \Theta_x^S & \Theta_{xy}^S & \Theta_{xz}^S \\ \Theta_{yx}^S & \Theta_y^S & \Theta_{yz}^S \\ \Theta_{zx}^S & \Theta_{zy}^S & \Theta_z^S \end{bmatrix} \tag{6.5}$$

The principal moments of inertia, and on the diagonal (principal axis of inertia), must all
be greater than zero by definition. If there are negative values or zero, there is an error. In
addition, you can also test the so-called triangle inequality, because it must apply:

---

[9]For those who did not attend any lectures on dynamics: STEINER's theory describes the conversion of
the mass moment of inertia from the center of gravity to any other point on the body or in reverse.

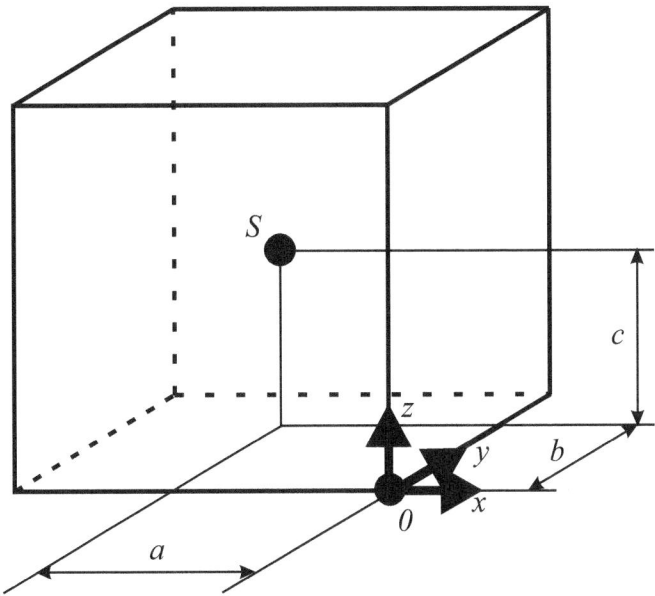

Fig. 6.10 Conversion with STEINER's theorem

$$\Theta_x^S + \Theta_y^S \geq \Theta_z^S \tag{6.6}$$

$$\Theta_x^S + \Theta_z^S \geq \Theta_y^S \tag{6.7}$$

$$\Theta_y^S + \Theta_z^S \geq \Theta_x^S \tag{6.8}$$

The deviation moments can be positive, negative or zero, so you cannot immediately see if they contain correct values. After all, you only have to calculate three, because the tensor is symmetrical and it applies:

$$\Theta_{xy}^S = \Theta_{yx}^S \tag{6.9}$$

$$\Theta_{xz}^S = \Theta_{zx}^S \tag{6.10}$$

$$\Theta_{yz}^S = \Theta_{zy}^S \tag{6.11}$$

Especially with mass properties, it is important not to use the standard values of the calculation program (not even those from the CAD program). If, for example, the mass and main moments of inertia are each given as 1 kg or 1 kg m², a tie rod (in the sense of technical mechanics a slim beam) would become a cube with at least approximately correct mass. If we take the internal combustion engine, a larger engine can still be approximately cube-shaped, but the mass is understandably two orders of magnitude higher. Why am I mentioning these *trivial examples* which you would never parameterize like this? Because it is happening! The mass data are difficult to obtain for some components and entering the correct values is postponed until later and then forgotten.[10]

6.2   Pre-Simulation Phase

6.2.1   Consistency of Data and Model

If several people are working on the same vehicle project, at different times or with different questions, they will usually do so at different workstations. A crucial question is, how is it ensured that they work with the same data and models? For speed reasons working on the local hard disk is usually the preferred way. Data access is much faster here than when working over a network. The moment a file is copied to different computers, it is no longer the same file. The probability that changes to data or models will not be made in all copies can be assumed to be 100% in the longer term. The data status on the different computers is then no longer consistent, but reflects, for example, different development phases.

How can it be achieved that only consistent data and models are used? With consistent adherence to the single-source principle. All calculations must be based on the same set of data and models. Software technology offers several possibilities for this.

You can use a database in which the data set and the model are entered separately. Changes can be traced here and an older dataset can also be retrieved consciously, for example to make a comparison of different development releases. If necessary, users can be granted different rights. Depending on their status, they can only retrieve or change the data and models.[11] Of course, it must still be possible to work with local variants. After all, development releases first have to be worked out. But as soon as they are *official* and are no longer modified, they must be protected from modification.

The version control systems from software development are based on a similar principle. If you are working on larger software projects, you need to be able to track changes, and since you are working with source code, differences between two versions can easily be displayed. The same mechanisms can also be used for the simulation data and the models. Version control is possible in any case. A data comparison depends on the selected data format and whether appropriate tools are offered for it. The models depend on whether they are saved in binary or text format.

---

[10]Mass parameters are often given, but the center of gravity position or the inertia are often more difficult to get. Even in the CAD-programs they are not imperatively rightfully given, since especially with acquisitional parts often only the outer shape is delivered.

[11]Depending on the organization structure, one can distinguish here between whether just the data should be adjustable (user) or the models as well (specialist).

The aim of both approaches is in any case that the user can simply retrieve a data status and be sure that he is calculating with the desired data. Large models work with a variety of model parameters. If you modularize them, the data may be spread over many different files. A quick manual check as to whether the data status is current and consistent is usually not possible. The principle of *hope* is not sufficient for a reliable calculation process.

6.2.2   Model Diversity

In addition to the consistency described above, this procedure offers another advantage. Today's vehicle models are often available in a wide variety of body variants. They are available as two-door, four-door or extended limousines, as station wagons or convertibles. In addition to the in-house standard type of drive, there may also be an all-wheel or hybrid drive. Depending on the motorization, there are different brake and/or axle concepts. The explosion of the variants, which should also cover the last niche, is generally deplored. But nevertheless they have to be tested and this is certainly a big advantage of simulation.

Usually, these variants are based on a basic version with which they share the majority of the parameters.[12] Differences will only occur in selected components. The computational engineer of a variant may have nothing to do with the creation of the basic version. If one were to create a separate complete data set for each variant, many parameters would be duplicated. The consequence is the same as before. If a parameter used by all variants changes, it must be maintained in all data records. Here, too, it makes sense to retrieve the data at the moment of the calculation from a common central location. The effort of data maintenance for the individual user decreases and confidence in the consistency of the parameters and thus in the result of the calculation increases.[13]

6.2.3   Simulation History

In the course of the project one will always want or have to show the hopeful progress of the stage of development. For this reason, these levels have to be *frozen*, for example, when a defined process milestone is reached. This means that from this point on, the data in this phase is no longer changed. In the next phase, work continues with a copy of the previous status and only this is then changed. That makes it easy to compare development statuses. In addition, you have a relapse level if the product develops in the wrong direction.

If decisions are made in the development process on the basis of the simulation results, one may have to take responsibility for one's statements later, when the product actually

---

[12]This, of course, is the chassis perspective on these variations.

[13]The effort, to develop such a consistent simulation process, is not to be underestimated. However, it is worth it or compulsory, if one works in the simulation environment described in this chapter.

exists. Maybe at this point you are already working on another project and no longer have the details in mind. Then it is imperative to be able to repeat the calculation and get an idea of the input data used at that time. See Sect. 6.4 for more information.

6.3 Simulation Phase

6.3.1 Local or Distributed

For the design of the actual simulation process, it strongly depends on whether the calculation is carried out on your own computer or on central calculation nodes. With today's multiprocessor systems, the workstation computer can also be used for single calculations, without which it is no longer available for anything else. If, however, many and even longer calculations are necessary, it seems more sensible to outsource the calculation process to external processors. It is not advisable to block the workplace, especially for calculations that take several hours.

6.3.2 Copying Procedure

With the local calculation, there is no need to copy the files necessary for the calculation.[14] Typically, there are one or more input directories and one output directory in which the calculation results are written. If, for example, you have a four-processor system, a maximum of three calculations can be performed simultaneously if only one processor is left for the use of the computer.

  If the simulation is used productively, there are usually many calculations to be carried out. For variant calculations alone, the number of necessary simulation runs can quickly increase.[15] A copy process is then required, which, depending on the operating system and the network environment, ensures that all required files are available at the beginning of the calculation. Here, it is natural to consider the concept of a database or a central storage of binding data and model sets with possible local working versions discussed in the last section. After successful completion of the calculation or after a termination due to an error condition, the generated results and log files must then be copied back to the workstation and the files on the calculation node deleted.

---

[14]Except one fetches them from an online database.

[15]Not to be confused with the brute force method, in which unreflectingly many variations are calculated and one only knows the results afterwards and only then draws conclusions. Rather than thinking ahead and only conducting the sensible calculations. A mean saying summarizes this with "Intelligence $\times$ computing power = constant".

6.3.3   Licenses

Depending on the simulation environment used, a different number of licenses may be required for the local and distributed calculations. If the local calculation is not possible, a GUI license and a kernel license[16] for the calculation nodes is sufficient, if such distinctions are made. This should be taken into account in the acquisition and price negotiation of licenses.

6.4   Post-Simulation Phase

6.4.1   Documentation of the Simulation

If the simulation is used productively in the development process, a large number of calculation results will certainly be generated. The results usually consist of at least one file with the calculated time series and a log file, whether the calculation was successful or if not, for what reason it was aborted.[17] The scope of the calculation logging can be set for most programs.

If you use a simulation program for the first time or if the entire procedure is new for you, you should not be afraid to make a very extensive logging. Of course, only if you take a look at the log file after completing the calculation. This should also be done at the beginning of successful calculations, so that you get an impression of what it looks like in such a case and what content there is. If a certain routine is set up, the log files should still contain at least so much information that one can read off the success of the calculation or, if the calculation is aborted, find out the cause. If the calculation only takes a few seconds, you can dispense with all other information except for the success message, because in case of doubt the calculation can be repeated quickly with increased protocol content. This procedure is forbidden for longer calculations.

6.4.2   Archiving

For some questions, many variants are calculated in order to read out an improvement of the product from the overall set of results. This means that most of the results that have not

---

[16]By this the actual core of the simulation software is meant, in which the motion equations are solved and integrated.

[17]If the calculation process is mostly automatized, one must check whether the calculation was really successful and whether it was calculated until the end or whether just a part of the calculation is available. Some results will be understood better.

contributed to an improvement can easily be deleted after viewing the results.[18] Some results may need to be kept a little longer, for example until the end of a series of tests, before they can also be deleted.

And then there are the results that are used for product decisions. These calculations are invoked and the results communicated. The question is, what do you keep from these results? And above all, how do you keep them? Do you use an archiving system or do they lie in a project directory, well structured, so that you can find them again when you search for them? Does it really make sense to archive the results, or should one not make the simulation run repeatable and save the necessary files? Of course, there is no universal answer to these questions. But you can see which topics you should deal with. Maybe not on the first day of your computing career, but then with increasing success.[19]

6.4.3 Motivation for Documentation

The necessity of logging and archiving often only becomes apparent when you have to explain to your bosses or clients that you no longer know how you came to this statement back then and that, no matter what you do now, you do not come to the same conclusion. You can wait and see if this moment ever happens to you (had I mentioned the principle of *hope*) or proactively avert it.

Documenting what you do is generally not one of the engineers' favorite pastimes. However, it does not contribute to credibility if you cannot prove what you have actually calculated. That this procedure does not correspond to the popular keywords *quality management* or *process reliability* should be accessible to everyone.

6.5   Reproducibility of the Simulation Results

The great advantage of simulation is its reproducibility. A calculation can be repeated at any time and always delivers the same result. That is the common opinion and you can read it everywhere. I do not want to turn a blind eye to this argumentation, but it contains some assumptions that do not always apply.

Assumption 1: The user is able to leave his simulation environment and models unchanged over time, ensuring that the same calculation is actually performed.

---

[18]Except if the non-effect of the improvement shall be documented. This too can of course be a question.

[19]From the point on, at which permanently more people than oneself are interested in the results.

So it depends very much on the care of the user whether his model is *alive* and is subject to ongoing changes. If a model error is found, it must be corrected. If a more efficient approach is found, it must be implemented to save time. There is nothing to be said against this if the old model exists unchanged and can be resuscitated again. Strictly speaking, this also applies to the simulation environment, the operating system and the entire simulation process.

Assumption 2: The user knows which parameters he actually calculated with and has logged them so that the same parameters can be used for the repetition.

The first two assumptions sound as if I do not trust the users of simulation programs to be careful. Of course, that is not the case. But I trust them to have to solve a current problem in the concrete task, sometimes to forget the future repetition that may only be required under certain circumstances, and thus to neglect the unpopular topic of documentation. Only a structured simulation process with a high degree of process automation can help here.

Assumption 3: The model is deterministic in its structure, i.e. its behavior is predetermined by clear rules. One can calculate the future behavior and thus each repetition leads to the desired same result.

In contrast to deterministic systems, stochastic systems can only be calculated with the help of probabilities. So there are random influences that cannot be predicted exactly. The easiest way to explain this behavior is to use the example of hardware-in-the-loop simulation, which is described in more detail in Chap. 17. If a field bus system such as the CAN (*Controller Area Network*) is integrated into the system, the transit times of messages can vary depending on the load on the bus. Depending on the integration step, the expected message may still be available on this or perhaps on the next time step. Real electrical or hydraulic components can be dynamic in different ways depending on the temperature. It is therefore of influence whether the entire system has just been switched on and has a constant room temperature or whether various calculations have already been carried out and some components have significantly higher temperatures.

Both examples are not really stochastic, since the running time of the message can be represented as a function of the bus load and the dynamic behavior of a hydraulic cylinder above the oil temperature. However, if these effects are not included in the modeled system and are considered correctly, they appear randomly.

What is the consequence of these statements? Of course, the assumption applies that the simulation itself is a reproducible tool—if used correctly.

References

[BaGK08]   BAUER, S., GRUBER, K. AND KNAUST, U.: CAD-/CAE-Integration - Entwicklung einer
             durchgängigen Berechnungsprozesskette, in *Berechnung und Simulation im
             Fahrzeugbau*, VDI-Berichte 2031, 201-216, 2008

[BrDG06]   BREITLING, T., DRAGON, L. AND GROßMANN, T.: Digitale Prototypen: ein weiterer
             Meilenstein zur Verbesserung der Abläufe und Zusammenarbeit in der
             PKW-Entwicklung, in *Berechnung und Simulation im Fahrzeugbau*, VDI-Berichte
             1967, 315-327, 2006

[DIN70000]   DIN 70000: *Straßenfahrzeuge, Fahrzeugdynamik und Fahrzeugverhalten, Begriffe*,
             1994

[Pohl10]   POHL, TH.: CAE-Mastermodels – the cornerstone of synchronised structural virtual
             engineering in a global environment, in *Berechnung und Simulation im Fahrzeugbau*,
             VDI-Berichte 2107, 45 – 61, Baden-Baden, 2010

[Rösk12]   RÖSKI, K.: *Eine Methode zur simulationsbasierten Grundauslegung von
             PKW-Fahrwerken mit Vertiefung der Betrachtungen zum Fahrkomfort*, Dissertation,
             University of Munich, 2012

[SAEJ670e]   SAE Recommended Practice SAE J670e JUL76: *Vehicle Dynamics Terminology*.
             SAE Handbook 3:34.447,Warrendale PA., 2003

Due to the multitude of applications and questions, there are of course no generally applicable recommendations or even a uniform procedure for simulation in chassis technology. For this reason, the models described below only have a classifying character. However, they should help to point out, which aspects should be considered and at which points one can pass (Fig. 1).

For applications in vehicle dynamics, ride comfort and load data prediction, proposals are made for implementation, which mainly originate from my own experience. As so often, there are almost always several ways to get there. There are approaches and tasks, which will weigh the treated aspects differently. Ultimately, everyone has to find and evaluate the right modeling for their own problem—but these explanations can be the basis for it.

The following chapters are now dedicated to the individual assemblies that can be found in the chassis. In addition, individual aspects that are necessary for the overall vehicle simulation are also considered.

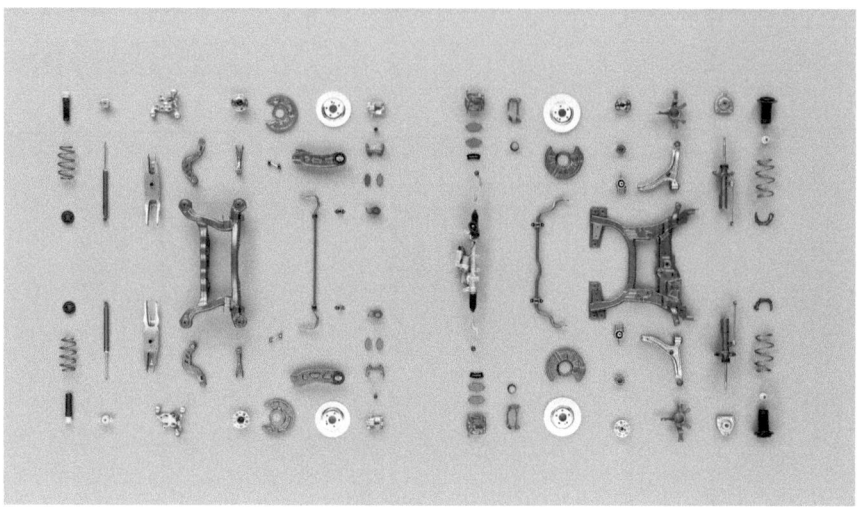

[Daimler AG]

Fig. 1 Suspension of a Mercedes B-Class (W245)

Modeling of Chassis Components

7.1   Fields of Application and Limits of Simulation

"The role of simulation in vehicle development is becoming increasingly important." Since I have been dealing with the topic of simulation—and that has been more than 30 years now—almost every second publication in this context starts with these or similar words. Accordingly, there should no longer be any other fields of activity in vehicle development. Of course this is not the case, but in fact the responsibility for the calculation has increased drastically during this time—this certainly does not apply to the budget of the computational engineers to the same extent. In every development process in the automotive industry, it is now at least as a so-called *Digital* or *Virtual Phase* firmly anchored.

The argument that simulations can replace drive tests and thus make the development of a vehicle faster and cheaper is often used—above all for economic reasons. This may be true for one or the other investigation even in this completeness, but in general I do not foresee the complete replacement of drive tests by simulations. As early as 1974, this was described in [Wieg74] as follows: "Simulation calculations are therefore already appropriate if they can be used to reduce the number of drive tests". The sensible use supports the drive test with good basic set-ups, which enable the test to deal only with the fine tuning and the associated customer suitability.

In the following, the three main topics for the simulation of chassis are briefly described with their respective modeling requirements, each with references to the corresponding component chapters. This is followed by a brief addition to the topic of modeling in the context of chassis simulation.

© Springer Fachmedien Wiesbaden GmbH, part of Springer Nature 2021   113
D. Adamski, *Simulation in Chassis Technology*,
https://doi.org/10.1007/978-3-658-30678-6_7

7.1.1 Vehicle Dynamics and Driver Assistance Systems

For the investigation of the dynamics of a vehicle, the mapping of the tire behavior under large longitudinal and transverse forces, up to the physical limit range, is essential. If the transfer behavior of the tire is not sufficiently mapped here, a completely different driving behavior will result. In the simulation, vehicle dynamics investigations tend to take place on ideally even road surfaces, so that the tire model must predominantly transmit long-wave road changes in order to be able to map curbs or bumps for driving maneuvers, for example (see Chap. 12).

In addition, it must be possible to reproduce the rigid body movement of the vehicle body so that the roll, pitch and yaw behavior can be investigated (see Chap. 15). Even if the vehicle dynamics are primarily determined by the longitudinal and lateral dynamics, the spring and damper behavior in the vertical direction must be mapped well in addition to the roll support by the anti-roll bar (and the position of the roll axis), because this also has an influence on the roll and pitch inclination of a vehicle (see Chaps. 9 and 10). Typically, one speaks of a relevant frequency range of up to 5 Hz.

Due to the low requirement for the frequency range to be investigated, even very simple chassis and vehicle models are suitable for vehicle dynamics investigations (see Chap. 8). For this reason, the models used are often real-time capable and this is extremely helpful for the development and design of driver assistance systems. Not only is less computing time required, but the possibility of using hardware-in-the-loop simulations to test the operability of these systems or to integrate real hardware into the simulation helps (see Chap. 17 and Sect. 7.1.4).

7.1.2 Ride Comfort

Driving maneuvers relevant to ride comfort do not take place in the physical limit range, so that the requirements for the horizontal dynamics of the vehicle and in particular of the tire are not very high. Nevertheless, curves are also taken, so that the rolling behavior and the transfer of lateral forces to the tire must be adequately mapped. A greater emphasis must be placed on the transmission paths of vertical dynamics. From tire-road contact to the kinematics of the wheel suspension and its connection to the vehicle body by means of elastomer bearings (see Chap. 8), suspension (see Chap. 9) and damping (see Chap. 10), from the vehicle body (see Chap. 15) to the driver's seat,[1] the transmission of forces and accelerations in the necessary frequency range must be ensured. Although the behavior of

---

[1]The path often ends at the driver's seat console, as acceleration sensors are usually installed here during the test. The seat itself has a very complex transmission behavior.

Fig. 7.1 Human perception of vibrations according to [HeEG13]

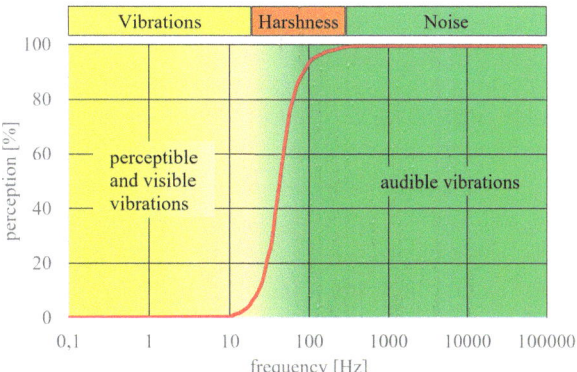

the tire under longitudinal and lateral forces can be simplified, the scanning of the road in z-direction is essential. The investigation of the shake[2] or micro shake behavior of a vehicle requires the correct mapping of the influence of the macro and micro roughness of the road surface (see Chap. 12).

If the steering train is also to be examined as a transmission element for road and unbalanced vibrations, the path from the tire to the steering wheel must be mapped accordingly (see Chap. 11).

In addition to the movement of the entire vehicle body, the transmission of vibrations beyond the rigid body movement is of particular interest. The method of multi-body systems limits the frequency range, but the detailed modeling of bearing stiffness and under certain circumstances of the first body shapes allow statements up to 30 Hz. In this way, investigations can be carried out in the area of visible and perceptible vibrations (Fig. 7.1).

If maneuvers are planned which involve larger wheel travel, the non-linear behavior of the chassis must also be taken into account due to large bearing deflections and spring progressions.

The requirements described lead to complex chassis and full vehicle models that are usually no longer capable of real-time operation. Nevertheless there is also the possibility to experience the results of the ride comfort simulation. In [AdJD07] a simulator is described, which can import results from the simulation and measurements of real vehicles and thus, before the first prototype is built, enables a vehicle evaluation regarding the vibration comfort. This is an excellent possibility to validate the complex simulation models subjectively.

---

[2]In the sense of SAE J670e, Vehicle Dynamics Terminology: "SHAKE – the intermediate frequency (5–25 Hz) vibrations of the sprung mass as a flexible body."

7.1.3 Load Data Prediction

If load signals are to be generated for the calculation of the durability of the chassis components or for the control of test benches, the requirements for model complexity are similar to those for ride comfort simulation. Only longer bearing travel or higher damper speeds are to be expected. In this case in particular, it is necessary to take a closer look at the component parameters and component characteristics to see whether this area is sufficiently mapped. It should also be noted that the elastomeric bearings are softer for large amplitudes than for smaller amplitudes of ride comfort.

Ultimately, however, a greater effort has to be made than the comparison of rainflow analyses or classifications of forces. These are very integral quantities that may conceal errors and inaccuracies in the model or measurement. Only a comparison in the time and/or frequency domain provides meaningful insights into model quality.

The tire models used for the ride comfort simulation are usually also able to transfer the excitations of a bad road section, but here, too, it is important to pay attention to the significantly larger amplitudes during parameterization. If the tire even punctures the rim, as can occur in some cases of misuse, this behavior must be covered by the model, because this goes far beyond a comfort load case.

7.1.4 Use of Simulators

A popular argument against the use of simulation in chassis development is the lack of subjective perception. No one should seriously want to replace the experienced test driver with a driver model in such disciplines. So far, human subjectivity cannot be comprehensively captured in algorithms—and hopefully it will remain so. Nevertheless, valuable support can be provided by suitable models in objectifying subjective impressions. Especially the simulators, which have become more and more powerful in recent years and in which one can actually experience the simulation results, offer a great—albeit not very inexpensive—possibility for this. These simulators can be used for training or the development of new vehicle concepts. Particularly in the field of driver assistance systems, there is a good opportunity for a customer acceptance analysis in advance. Large simulators in which a complete vehicle can be accommodated are operated by Daimler AG in Sindelfingen, Germany [Depp10] (Fig. 7.2), Toyota Motor Corporation in Shizuoka, Japan [MTS09] or the Forschungsinstitut für Kraftfahrwesen und Fahrzeugmotoren in Stuttgart, Germany [FKFS09].

The Brockhaus [Broc14] describes a simulator as: "A device or system with which certain behaviors, properties, and so on of a physical, technical, cybernetic, or abstract system as well as process sequences can be represented or imitated; especially a device with which (technical) conditions are created that correspond to those of devices to be handled later". The conditions to be created are an essential point. Consistency must be ensured, i.e. the behavior of the simulator must be plausible for the user and largely

[Daimler AG]

Fig. 7.2 Driving simulator of the Daimler AG in Sindelfingen, Germany

correspond to the behavior of a real vehicle within the scope of the investigations. If the behavior is artificial and alien, one will usually not be able to achieve satisfactory results. For this reason, it is important that the time behavior of the model and the implementation in the simulator are consistent with the expected behavior. If there are delays, steps or too low or too high amplitudes, the impression of a *toy* occurs and the user does not behave as in a real vehicle. From my own experience I know that in a consistent simulator you look into the rear-view mirror when you change lanes—although you know that this *only* is a simulation.

The reward for the not inconsiderable effort is that you do not have to model the driver in his complex behavior patterns and that you can try out difficult or dangerous driving situations with normal drivers without harming anyone.

7.1.5 Potential of Calculation or Undiscovered Treasures

Models are often created to map existing or future chassis components or chassis systems and then later compare the calculation results with real measurements. It is therefore important to map the behavior of the component or system as accurately as possible.

The opposite way is used less often. Not the component is modeled, but the desired behavior. This desirable behavior is described with a model approach that is as simple as possible. If the vehicle model has the required properties, consideration is given to which real component could achieve this behavior and how the component would have to be tuned. In this way, it is possible to come up with system solutions that nobody had thought of before. However, one now has at least one reference of a theoretical boundary case which cannot be reached by real systems. But you can show how close you can get to this limit depending on your system selection. For this, however, it is essential to move away from the orientation on the component and keep an eye on the overall vehicle behavior.

Of course, it is just as important to have sufficient component and system experience to be able to check suitable implementations.

7.2   Complexity of Models

If you have to deal with different questions that require different levels of detailed modeling, you can go at least two ways. First, there is the dogma that a model should always be as accurate and detailed as the question requires. An approach that I very much welcome, because why using a sledge-hammer to crack a nut? Example: If I want to determine the natural lift frequency of a vehicle to the nearest tenth of a Hertz, a single or dual mass oscillator is sufficient as a model. A model with 60 degrees of freedom would be clearly oversized. If, however, I want to determine the load on the vehicle body at the anti-roll bar bearing during a bad road crossing, I cannot avoid a detailed vehicle model with a complex tire model. From these examples it can be concluded that an optimally adapted or at least a scalable model is needed for every problem.

The other approach seeks to avoid precisely this. It assumes that if you need the most detailed model for the most complex problem, then this model should also be able to capture the simpler load cases. However, this approach often leads to the false conclusion that a model has now been generated for all possible load cases.

Both approaches are justified and we should briefly consider the advantages and disadvantages of these two methods.

7.2.1   Maintenance and Modifications

A separate model for each complexity level

- Any change may need to be made in each of the models.
- The great danger is that not every model is up to date and different versions are formed very quickly.[3]

---

[3]The discipline required to make the change conscientiously in all derivatives is high. To hold out independently of telephone rings, boss inquiries, lunch breaks and other distractions is heroic and thus reserved for very few of us.

One model for all complexity levels

- A change is made once and then applies to all models.
- An exception is when, for example, logical switches are used within the model to switch between simple and complex modeling.[4]

7.2.2 Computing Time Requirement

A separate model for each complexity level

- Depending on the model complexity, the computing time will vary, so that the simple models, for example, can be real-time capable and the complex ones often no longer.

One model for all complexity levels

- If the model is no longer real-time capable due to its complexity, this usually also applies to *simple* load cases.

7.2.3 Parameter Requirements

A separate model for each complexity level

- The basic parameters of a chassis are available relatively early or can at least be estimated well. Thus, calculations with the simpler models are possible very early in the development phase. For the more complex models, it is then necessary to wait longer until, for example, the measurements of the target components are available.

One model for all complexity levels

- With few exceptions, even the simplest load cases can only be calculated once the model has been fully parameterized. This means that a great deal of data and this data must be available in the required quality. Usually, it is only possible to start the first calculations relatively late.[5]

---

[4]This procedure can often be useful to save computing time. However, if you have to make this change manually, there is a great potential for errors. Here automation via a higher-level load case control is an option.

[5]This is practically a no-go for management enquiries in the concept phase. Here you need a short response time and you need to be able to achieve useful results with little information. In this regard, however, please refer to Chap. 6 and the GIGO concept.

7.3   Simple Model Approaches

Depending on the problem, simple (but not necessarily) linear models can usually be used. In particular, the single-track model popular for transverse dynamics or the dual-mass oscillator model for vertical dynamics should be mentioned. These models are sufficiently described in the literature (e.g. [MiWa04] and [Zomo91]) and will not be discussed further here. These simple model approaches often date back to a time when it was necessary to be able to solve the model equations on paper. Before the computers and the calculation programs found their distribution, the equations had to be solved by hand and accordingly closed solvable approaches were preferred. An excellent example of such a complete vehicle model for a passenger car and a truck can be found in [MiKB95]. There also all necessary equations and system matrices are to understand. Now one might think that with today's availability of high-performance computers and the most complex calculation programs, there is no longer any need for these models. It should be noted that a fully understood model with 2 degrees of freedom is usually more valuable to use than 1 with 100 degrees of freedom, in which one can only guess what connections there are in it. They are particularly important for a basic understanding of the effects of individual parameters on vehicle behavior.[6] Of course, the areas of application and the informative value of such models are limited—but so is our capacity for abstraction. Already in 1989 [Fial89] recognized: "Today there are vehicle models with many degrees of freedom and arbitrarily accurate replication of all non-linear compliance properties. That makes every detail explainable, but it damages the clarity."

   In the following it is assumed that more complex models are used in an full vehicle model. Nevertheless, the described procedures can also be applied to partial aspects of these simple models.

7.4   Where Is the Right Information?

Usually, the chassis tuning serves to achieve a desired driving behavior or to stop an undesired one. It depends on the know-how of the vehicle tuner, which vehicle parameters he varies and with which maneuvers he checks the result and with which negative side effects he excludes or detects.

   In addition to the subjective impression, the result is often also recorded objectively, i.e. through a measurement. In a real vehicle there are a large number of parameters that can be measured to get an insight into the behavior of the vehicle. Which vehicle coefficients are to be evaluated is usually obvious to the professional. But especially for beginners or complex situations, this question cannot always be answered unambiguously.

---

[6]Anyone who has fully penetrated the model with 100 degrees of freedom and understood the interactions is of course welcome to use it and feel patted on the back.

In the simulation, the number of measurable quantities is many times higher. The question about the *right* coefficients is the same as for the measurements of real vehicles. The more complex the models and/or the questions become, the more confusing the situation can be for the computational engineer. Here, too, a great deal of know-how about driving behavior is required. In addition, many vehicle coefficients can be determined in the simulation that cannot be checked in a real measurement. Here is the question of how to check the plausibility of these variables, since they cannot be directly compared with measured values. In contrast to the real vehicle, the model, or even better the sub-model, can be loaded with test signals at significantly more inputs in order to check the transmission behavior. In this way, the model can be checked for plausibility (if the transmission behavior is known or can be estimated). If there is a faulty behavior, the fault can usually be located more easily than with a real vehicle.

7.5 Planning and Evaluation of Maneuvers

7.5.1 Settling Time

In the fewest maneuvers, the chassis model or the full vehicle model will be in a steady state vibration. This state represents the state of equilibrium of the vibrating system, i.e. the weight forces of all masses are in equilibrium with the spring forces supporting them. Depending on how complex the force offsets were determined in advance, the model will need a short settling time before it has sufficiently settled and its natural oscillation has thus subsided (Fig. 7.3).

Since the evaluation of maneuvers is usually not interested in the transient response, there are two ways of dealing with this behavior.

If the configuration of the model changes frequently so that a new equilibrium position has to be found each time, it is advisable not to determine the rest position in advance and to let the model oscillate during each simulation. The time required for this should be in the small one-digit seconds range (simulation time). Now you have to cut this part off from each maneuver and evaluate only the behavior according to it. For this, however, the relevant part of the maneuver (e.g. start of a road section) must also be placed at the position after the settling time. In order to avoid having to determine a new settling time every time, a constant interval is recommended, which must be so large that the settling process is in any case within the interval. If different speeds are used, this must be taken into account when designing the maneuver, as explained in the next section.

For the second possibility, many simulation programs offer the possibility of using the state vector of the last calculation step of a previous, saved simulation for the initial values of the state vector of the new calculation (Fig. 7.4). It must of course be ensured that the same boundary conditions prevail and that all maneuver-relevant variables are contained in this state vector.

Fig. 7.3 Settling phase of the entire vehicle

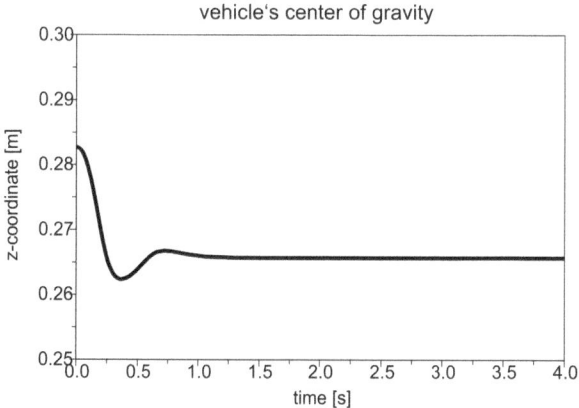

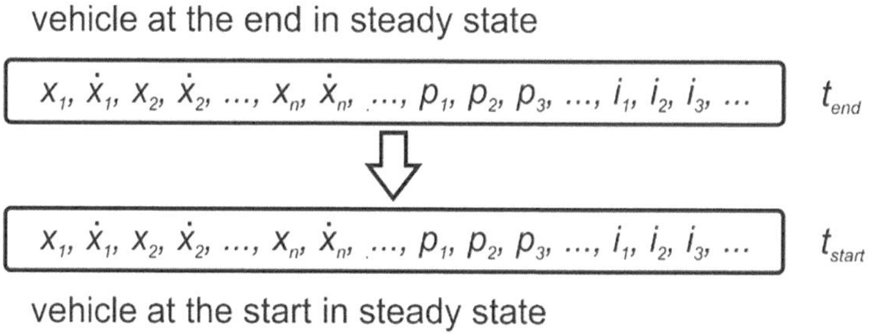

vehicle at the end in steady state

$$x_1, \dot{x}_1, x_2, \dot{x}_2, \ldots, x_n, \dot{x}_n, \ldots, p_1, p_2, p_3, \ldots, i_1, i_2, i_3, \ldots \qquad t_{end}$$

⇩

$$x_1, \dot{x}_1, x_2, \dot{x}_2, \ldots, x_n, \dot{x}_n, \ldots, p_1, p_2, p_3, \ldots, i_1, i_2, i_3, \ldots \qquad t_{start}$$

vehicle at the start in steady state

Fig. 7.4 Use of the state vector from a previous calculation

7.5.2 Length and Duration of the Maneuver

When planning a maneuver, it should be noted that the values of the vehicle speed (v [m/s]), the track length (S [m]) and the simulation time (t [s]) are directly linked by a simple relationship.

$$t = \frac{S}{v} \tag{7.1}$$

I would not mention this if it were not a popular error possibility. A given distance of constant length[7] is to be crossed by a vehicle at different speeds. In order to ensure that only the desired section of the route is crossed for each vehicle speed, the simulation time must of course be adjusted. Otherwise, depending on the program and road definition, simulation stops (good) or the consideration of possibly level runouts (bad) may occur.

---

[7]Not unusual on roads.

You can use different procedures to evaluate these maneuvers. In the case of integral evaluation quantities that evaluate the entire maneuver, one ultimately only has to ensure that the maneuver corresponds to the specifications.[8] Otherwise, at different speeds, an output of the quantities of interest over time is usually not very informative. A comparison over the trajectory is more appropriate here. When driving straight ahead this is *trivial* because only the size above the x-coordinate must be evaluated. If the trajectory is spatial, a path parameter must be defined that takes into account the distance travelled, comparable to the odometer in the vehicle. Some programs offer this by default, others require this parameter to be calculated.

References

[AdJD07] ADAMSKI, D., JUST, W. AND DRAGON, L.: Subjektive Bewertung des Ride Komforts von Digitalen Prototypen mit Hilfe eines Ride Simulators, in *Erprobung und Simulation in der Fahrzeugentwicklung*, VDI-Berichte 1990, 289-302, Wuerzburg, 2007

[Broc14] Die Brockhaus Enzyklopädie Online, F. A. Brockhaus/wissenmedia in der inmediaONE GmbH, Guetersloh/Munich, accessed on 17.02.2014

[Depp10] DEPPE, PH.: *Fahrsimulatoren im Mercedes-Benz Technology Center in Sindelfingen*, blog.mercedes-benz-passion.com, 2010

[Fial89] FIALA, E.: Reifen, Fahrer, Lenkverhalten, in *Reifen-Fahrwerk-Fahrbahn*, VDI-Berichte 778, Hannover, 397-423, 1989

[FKFS09] FKFS, *Der Stuttgarter Fahrsimulator*, brochure, Forschungsinstitut für Kraftfahrwesen und Fahrzeugmotoren, Stuttgart, 2009

[HeEG13] HEIßING, B., ERSOY, M. AND GIES, S. (ED.): *Fahrwerkhandbuch*, Vieweg+Teubner, Wuerzburg, 2013

[MiKB95] MITSCHKE, M., KLINGNER, B. AND BRAUN, H.: Zulässige Amplituden und Wellenlängen herausragender Unebenheitsanteile – Einfluß von Einzelhindernissen und Periodizitäten auf Fahrkomfort, Straßen-, Fahrzeug- und Ladegutbeanspruchung sowie Fahrsicherheit, Schriftenreihe *Forschung Straßenbau und Straßenverkehrstechnik*, Heft 710, Bundesministerium für Verkehr, Bonn, 1995

[MiWa04] MITSCHKE, M. AND WALLENTOWITZ, H.: *Dynamik der Kraftfahrzeuge*, Springer Verlag, Berlin, 2004

[MTS09] MTS: *Das Verhalten des Fahrers besser verstehen*, MTS customer case study, Eden Prairie, 2009

[Wieg74] WIEGNER, P.: *Über den Einfluß von Blockierverhinderern auf das Fahrverhalten von Personenkraftwagen bei Panikbremsungen*, Dissertation, University of Braunschweig, 1974

[Zomo91] ZOMOTOR, A.: *Fahrwerktechnik: Fahrverhalten*, Vogel Verlag, Wuerzburg, 1991

---

[8]Here we have to take into account what we have implemented, not what we wanted to implement.

In addition to carrying and guiding the vehicle, the task of the suspension is also to isolate it from disturbing influences such as road roughness, unevenness or wheel imbalances (Fig. 8.1). Depending on which of these aspects comes to the fore, the necessary complexity must be selected during modeling. First of all, it has to be clarified whether it should be a pure kinematical model, in which the simulation of the chassis bearings is omitted and only ideal joints[1] are used, or whether the compliance has to be considered. The differences are discussed below.[2]

The various wheel suspension concepts are only described as examples in the next section. If you want to get a closer look at the kinematics of the different suspension concepts, you should consult [Henk93], [Mats07] or [ReBe00].

8.1 Modeling of the Kinematics

8.1.1 Mechanism-Oriented Models

In a mechanism-oriented model, kinematics in the sense of rigid-body mechanics is represented by ideal joints and connecting elements. The GRÜBLER-KUTZBACH-criterion (Eq. 8.1) determines the number of degrees of freedom in a mechanism with n_B bodies and n_J joints with f_{Ji} degrees of freedom per joint. The parameter r describes the number of

---

[1]Ideal joints have no expansion, no mass, no elasticity and are friction-free.

[2]If there are no elastomeric bearings installed, as is sometimes the case in racing, this question naturally does not arise.

© Springer Fachmedien Wiesbaden GmbH, part of Springer Nature 2021
D. Adamski, *Simulation in Chassis Technology*,
https://doi.org/10.1007/978-3-658-30678-6_8

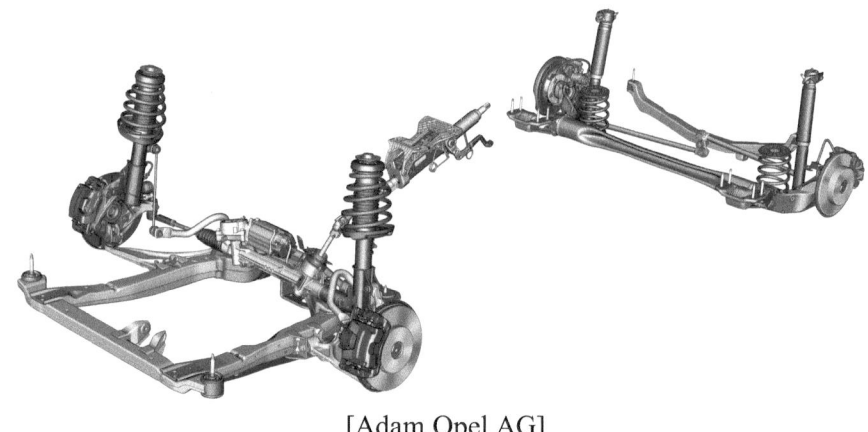

[Adam Opel AG]

Fig. 8.1 Front and rear axle of the Opel Astra H

so-called isolated degrees of freedom that do not contribute to the actual movement of the mechanism.

$$f = 6\,n_B - \sum_{i=1}^{n_J} \left(6 - f_{J_i}\right) - r \tag{8.1}$$

Table 8.1 shows a selection of relevant joints with the number of degrees of freedom bound by them.

If these standard joints are connected in a suitable way, some basic elements of suspension concepts are created to guide the wheel movement. The prismatic joint can be found, for example, in the rack-and-pinion steering system.

The unguided wheel carrier has six degrees of freedom as a rigid body, which must be reduced to one in the case of independent wheel suspension. In the following, kinematics calculations are carried out, meaning only joints are used. In the axle or the suspension of course also elastomeric bearings are installed, but this is only considered in the compliance.

A simple component is a link (Fig. 8.2), which consists of two ball joints and a rigid connection. With the help of the GRÜBLER-KUTZBACH-criterion, the number of degrees of freedom bound by the link can be calculated.

The first joint is connected to the so-called root or base, which is not counted as a body, in our case to the vehicle body. Thus the mechanism has two bodies—the link and the wheel carrier—and initially has $f = 2 \times 6 = 12$ degrees of freedom. Ball joints are attached to both ends of the link, each of which binds three degrees of freedom, leaving six degrees of freedom for the mechanism.

The wheel carrier would have had so many degrees of freedom even without the link, which is why it initially appears unnecessary. But now the isolated degree of freedom

Table 8.1 Schematic overview of typical joints

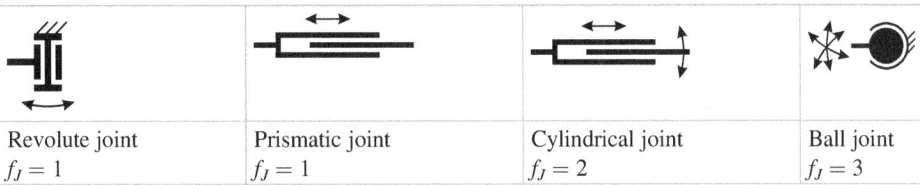

Revolute joint	Prismatic joint	Cylindrical joint	Ball joint
$f_J = 1$	$f_J = 1$	$f_J = 2$	$f_J = 3$

Fig. 8.2 Link   body side

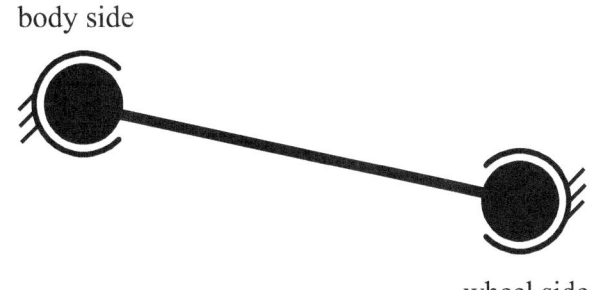

wheel side

comes into play. The link can rotate around the connecting axis of the two ball joints. However, this has no influence on the movement of the wheel carrier and is therefore deducted.[3]

Let us recapitulate:

Number of bodies:
   $n_B = 2$
Number of joints:
   $n_J = 2$
Degrees of freedom per joint:
   $f_{J1,2}$ = each 3
Number of isolated d.o.f.:
   $r = 1$
Number of total d.o.f.:
   $f = 2 \cdot 6 - (6 - 3) - (6 - 3) - 1 = 5$

The number of degrees of freedom of the wheel carrier was reduced from six to five by connecting the wheel carrier with a link.

If four further links are now attached, the number of degrees of freedom of the wheel carrier is limited to one, so that only the desired wheel travel remains. This axle concept is

---

[3]With the real axle, the isolated degree of freedom does not play a role, since, except in racing, an elastomer bearing is always used on the body side, so that the link cannot spin.

Fig. 8.3 Multi-link rear
suspension of the Mercedes
C-Class (W205)

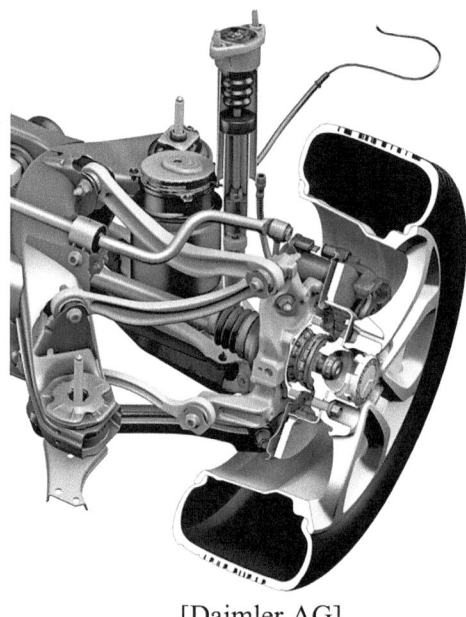

[Daimler AG]

known as a multi-link suspension (Fig. 8.3) and can be found in various derivatives at
various manufacturers on the rear axle.

Depending on the counting method, it can also be found as a four-link or as a five-link
suspension on the front axle (Fig. 8.4). The fifth link is then connected to the steering as a
tie rod.

If you connect two links by combining the ball joint on the wheel carrier side in one
point and replace the ball joints on the body side with revolute joints, you get an A-arm.[4]
Although two revolute joints can be seen in Fig. 8.5, they are only counted as one, because
they have the same axis of rotation and must therefore be considered kinematically like one
joint.

If we repeat the same calculation as before, we see that the A-arm binds two degrees of
freedom of the wheel carrier.

Number of bodies:
  $n_B = 2$
Number of joints:
  $n_J = 2$
Degrees of freedom per joint:
  $f_{J1} = 1$ (revolute joint), $f_{J2} = 3$ (ball joint)

---

[4]Also known as wishbone or control arm.

Fig. 8.4 Multi-link front
suspension of the Audi A6 (C7)

[Audi AG]

Fig. 8.5 A-arm

body side

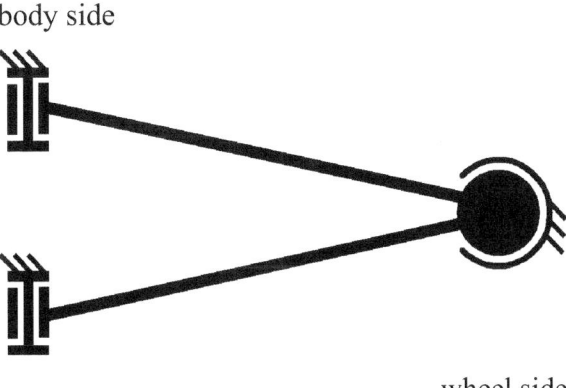

wheel side

Number of isolated d.o.f.:

$$r = 0$$

Number of total d.o.f.:

$$f = 2 \cdot 6 - (6 - 3) - (6 - 1) = 4$$

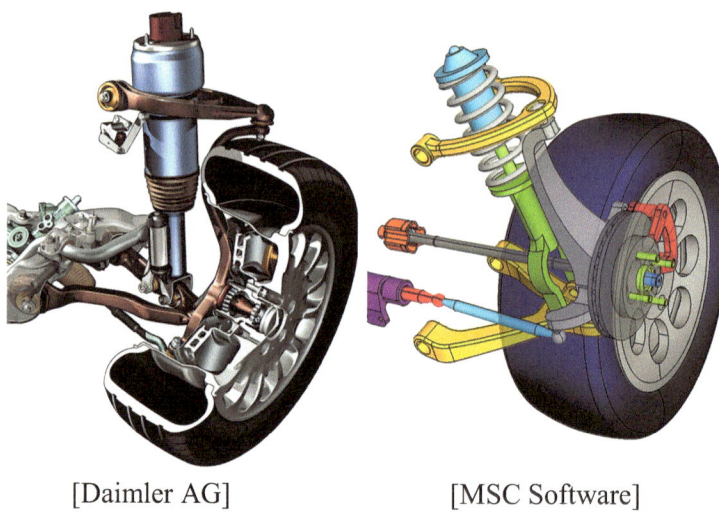

[Daimler AG]   [MSC Software]

Fig. 8.6 Double wishbone suspension, component and simulation model

Of the original six degrees of freedom, only four remain. If we combine two A-arms with one link, five degrees of freedom are bound and we have the basic shape of the double wishbone suspension (Fig. 8.6). If we connect the tie rod with the steering, the axle becomes steerable. If it is connected to the vehicle body or to the sub frame, as is usual on the rear axle, it is not steered.

The double wishbone suspension is available in various derivatives. If one dissolves an A-arm and replaces it with two links, one usually obtains advantages in installation space and expands the scope of the set-up. If one dismantles both triangular control arms, one gains the previously discussed five-link suspension.

A mechanism that can tie four degrees of freedom alone is the trapezoidal link (Fig. 8.7). It can be found on the so-called trapezoidal link suspension in combination with a link on the rear axle (Fig. 8.8).

The revolute joints on the vehicle body side are again to be regarded as one, kinematically. As a revolute joint is also installed at the wheel carrier side, this component binds considerably more degrees of freedom than the A-arm. The calculation of the degrees of freedom yields:

Number of bodies:
  $n_B = 2$
Number of joints:
  $n_J = 2$
Degrees of freedom per joint:
  $f_{J1,2} = $ je 1 (revolute joint)

Fig. 8.7 Trapezoidal link

wheel side

Fig. 8.8 Trapezoidal link suspension as rear axle of the Audi A6 (C7)

[Audi AG]

Number of isolated d.o.f.:
  $r = 0$
Number of total d.o.f.:
  $f = 2 \cdot 6 - 2 \cdot (6 - 1) = 2$

The last kinematic concept I would like to present is the world's most popular front axle—the McPherson axle (Fig. 8.9). It is a typical front axle for front-wheel-drive vehicles up to the upper middle class and, as this type of vehicle is the most frequently sold worldwide, it can almost be regarded as the standard in this segment.

Fig. 8.9 Hyper Strut front axle
of the Opel Cascada

[Adam Opel AG]

It consists of a lower control arm and a link, but only three degrees of freedom are bound by this. In contrast to the suspensions discussed so far, the damper not only functions as a force element, but also as a wheel travel. The damper is firmly connected to the wheel carrier and can thus transmit torques. Kinematically, it is to be regarded as a cylindrical joint that can bind two degrees of freedom. Since dampers are not suitable for the transmission of bending moments due to their basic task, a stronger piston rod must be used in addition to lateral force compensation by adjusting the mounting spring. If we also calculate the degrees of freedom here, the following picture results for the entire suspension:

Number of bodies:
  $n_B = 4$ (A-arm, link, wheel carrier, damper)
Number of joints:
  $n_J = 6$
Degrees of freedom per joint:
  $f_{J1\text{-}4}$ = each 3 (ball joint), $f_{J5} = 1$ (revolute joint), $F_{J6} = 2$ (cylindrical joint)
Number of isolated d.o.f.:
  $r = 2$ (link, piston rod of the damper)
Number of total d.o.f.:
  $f = 4 \cdot 6 - 4 \cdot (6 - 3) - (6 - 1) - (6 - 2) - 2 = 1$

The tie rod was also taken into account here, although it is of course connected to the steering on a front axle and thus realizes the degree of freedom for steering. If the A-arm is disassembled, i.e. replaced by two links, the so-called three-link suspension is obtained.

A good overview of the kinematics behind a selection of suspension concepts can be found in [ScHB10]. There it is described in detail which methods of higher mechanics can be used to solve kinematics with the help of the method of multibody systems.

8.1.2 Map-Oriented Models

For considerations, in the field of vehicle dynamics it is particularly relevant how the wheel moves under the external forces and moments, i.e. how the toe and camber angles change. This has a decisive influence on the driving behavior. In addition to the requirements of real-time capability, which is indispensable for hardware-in-the-loop simulations, the detailed models considered above reach their computing time limits. Now models are more in demand, which can represent the behavior sufficiently exactly, without necessarily having to know which components are responsible for it [Kune05]. The consideration of the individual stiffness of the chassis bearings leads to a numerically stiffer system, which ultimately increases the computing time [Tobo04]. We are therefore looking for an approach that allows the consideration of kinematics and compliance without explicitly modeling them.

For the approach of the map-oriented models, it is necessary that the kinematics and compliance of the suspension can be mapped either by measurements on the real chassis or, for example, by an MBS simulation with a complex suspension model. The wheel movement is stored separately in maps according to kinematics and compliance, (Fig. 8.10) and then merged to determine the wheel position and orientation by double-linear interpolation [Lang97].

If no variables are to be determined by differentiation during the calculations, the velocity states must also be stored in characteristic maps. To generate the maps, the suspensions are fully compressed and deflected. Toe, camber, spreading axis, etc. are recorded and stored in the characteristic maps. For steerable axles, additional steering from stop to stop is required.

Since the individual links of the suspension are not modeled, their masses cannot directly be taken into account. It is advisable to add half of this mass proportion on the vehicle body side and half on the wheel carrier side. Force elements such as springs, dampers or anti-roll bars must be converted in their force application to the wheel carrier, since the actual points of application are not available (see also Sect. 9.4).

It is obvious that the evaluation of maps is much faster than the complete calculation of kinematics and compliance. A statement about the acting forces in the bearings or about the influence of a certain bearing or joint is not directly readable, since they were not explicitly modeled. In the case of changes in the chassis, the influence of the change can only be shown after the creation of new maps and the new calculation.

Commercial simulation programs that work with map models often offer the possibility to generate the necessary maps from a kinematic model of a suspension by preconfigured load cases in a more or less simple way. This is basically possible with any MBS program

Fig. 8.10 Map-oriented model
of a suspension

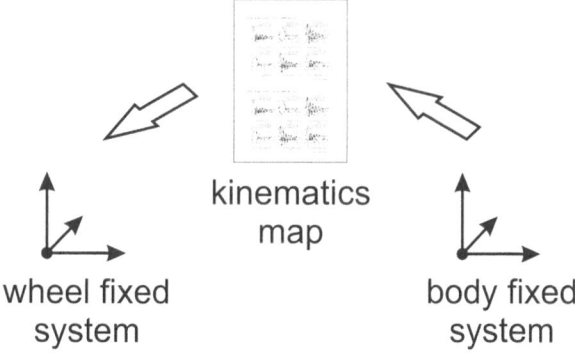

kinematics
map

wheel fixed
system

body fixed
system

in which the suspension can be mapped, but such a program is not always available or there
is the necessary knowledge of how to operate it. Figure 8.11 shows a possibility how to use
the simulation environment CARMAKER using IPG KINEMATICS maps from a kinematics and
that you have a graphical control possibility to check whether the kinematics have been
adopted correctly.

8.1.3 Behavior-Oriented Models

If an existing real vehicle is to be modeled for which the necessary kinematic data is not
available, another way must be found to map the wheel movement. A complex three-
dimensional measurement of the kinematic points and the measurement of all bearings is
usually too time-consuming, too expensive and only makes sense in very few cases. The
class of behavior-oriented models is based on the observation of wheel motion under
different load cases, in order to transfer this behavior to a simple basic model. The
parameters of the basic model must be identified, i.e. they must be adapted until the
behavior of the model and the suspension sufficiently match. This procedure has already
been presented in Sect. 3.4.2.

In [HaHo03], this process of identifying the dynamic behavior of the suspension using
the method of neural networks is described. Basic methods for identifying parameters of a
measured system are, for example, listed in [Ise192].

8.2   Modeling of the Compliance

8.2.1 Elastic Chassis Parts

This section deals neither with the springs (see Chap. 9) nor with the elastic bearings (see
Sect. 8.3). This is about the elastic sheet metal structures that occur, for example, with the
twist beam or sword-link suspension or a sub frame. However, elasticities can also occur in

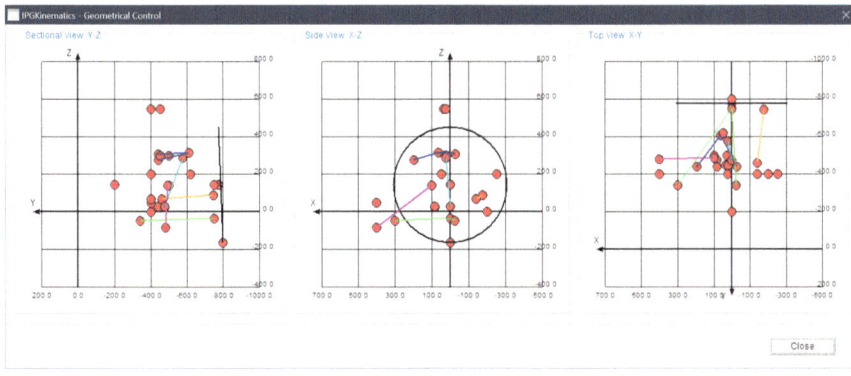

[IPG Automotive]

Fig. 8.11 Kinematic control of a suspension with IPG Kinematics

wheel carriers, such as a double wishbone front axle (long spindle) which have an influence on kinematics and compliance (Fig. 8.6).

Almost all vehicles today are subject to extreme weight targets, which can only be achieved through corresponding lightweight construction concepts. Beyond that, it is still required to get the Nordschleife ability, speak a *sporty* set-up, so the stiffness of the force transmission points in the vehicle body and the bearing stiffness are no longer separated by orders of magnitude, as was previously demanded. This results in measurable displacements with large lateral or longitudinal forces. If an elastic body model is not used (see Chap. 15), a *secondary spring rate* should be considered elsewhere. In the following, suggestions are made for the modeling of these effects.

A simple approach to model elastic chassis parts using the MBS simulation is to model the component as rigid and then represent the elastic parts using spring elements. Using the example of the twist beam axle (Fig. 8.12), the trailing arms could be rigidly attached to the cross member and the cross member could be modeled from two rigid parts connected by a revolute joint and a superimposed torsion spring. The torsional stiffness required for this must then be determined from tests or FEA calculations.[5] That way, the global deformation behavior can be mapped. Stresses or local deformations cannot of course be dissipated in this way.

This would require more elaborate beam models, which would, however, require considerably more computing time and parameterization. In [LiSe01] a model is described which was obtained from a FE model by means of modal reduction.

In the case of the wheel carrier, one occasionally finds the implementation that the transverse stiffness is not accommodated in the component wheel carrier, but in the

---

[5]Even if metals have a negligible material damping, no undamped oscillator should be modeled. Due to the high stiffness a high frequency interference would occur, which at least slows down the integration process.

Fig. 8.12 Twist beam axle

[Audi AG]

connection to the links. This has the advantage of a relatively simple modeling of the wheel carrier—as a classic rigid body. However, depending on the type of modeling, this violates the following modeling requirement:

If the parameters of a concrete component are known, they must also be used in the model.

The superimposition of the component stiffness of the wheel carrier would lead to a falsification of the bearing data. If the bearings are not explicitly modeled and all secondary spring rates are combined in an imaginary spring, then of course the wheel carrier stiffness can also be taken into account here.

Another approach is to model the ball joint between wheel carrier and link not ideally rigid, but elastic.[6] However, the same modeling requirement applies here. If the joint stiffness is based on a measurement, this must be taken into account in the joint model and is therefore not available as a tuning parameter.

The required stiffness can be determined either from a component measurement or an FE calculation. Most MBS simulation packages contain approaches that allow elastic beam models to be used. However, the computing time and the numerical stability must be taken into account.

Finally, a coupling with an FEA program can also be used, in which the elastic structure was modeled out and which provides a suitable modal description for the MBS program. One must consider here whether this approach still covers the actually occurring, possibly non-linear, displacements by the model. If only linear relationships are considered here, only vibration phenomena with correspondingly small amplitudes can be sensibly investigated.

---

[6]Joint models are used mainly because of friction. Both effects, the high stiffness and the transition between static and sliding friction, slow down the numerical analysis and lead to high computing times due to small step sizes.

The fact that even the global torsional stiffness of the vehicle body can be represented by simple models for some questions is described, for example, in [Kroh01].

8.2.2 Secondary Spring Rate

The term secondary spring rate includes all stiffness that occur from the wheel to the vehicle body and that do not belong to the suspension spring or tire. This refers to all elasticities in the bearings, the links and, if necessary, in the sub frame or in the force transmission points of the body.

The following Sect. 8.3 describes how the bearing elasticities can be represented in different complexity in relation to components. This approach requires that the position, orientation, stiffness and damping are known for each bearing. Only then can you model in a component-oriented way. The advantage of this method is that the bearing travel and the forces in the bearing and at the point of transmission into the body can be calculated. However, this information is not always available or the model complexity was chosen so that no explicit components were modeled (as for example with the map-oriented procedure from Sect. 8.1.2).

The secondary spring rate can also be interpreted as an additional spring connected in parallel to the suspension spring and can be modeled accordingly (Fig. 8.13). If you now measure the total spring rate of a real vehicle $c_{total} = c_{spring} + c_{secspring}$ on a K&C test bench the spring stiffness of the suspension spring c_{spring} must be measured separately, i.e. when removed, and can then parameterize both model parts. With this approach, each modeled component must be parameterized with its own stiffness. Only the remaining, unknown stiffness can then be added to the secondary spring rate. If a bump stop is modeled, then the corresponding stiffness must be found in the bump stop model, with a damper the gas force must be found in the damper model.

This is a linear approach that will be sufficient for many applications. However, the effects on the individual bearings are not taken into account and cannot be evaluated accordingly. Still, the lack of damping is much more serious. Elastomer bearings are used for vibration isolation due to their material damping. For simulations in the ride comfort environment, this type of modeling is therefore largely out of the question.

8.3 Simple Elastomeric Bearing Models

In contrast to an ideal joint, which only allows the respective degrees of freedom of movement and blocks all other movements,[7] an elastomer bearing must allow movements in all directions under the influence of force or torque. In most cases, the model type used is

---

[7]More correctly, kinematic constraints are defined.

Fig. 8.13 Distribution of
suspension spring stiffness and
secondary spring rate

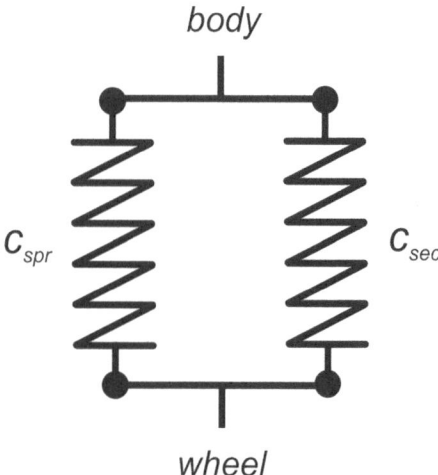

the force element *bushing* which, like all force elements, has neither an expansion nor a mass (Fig. 8.14).

Let us first look at an example of a force direction. Depending on a displacement dx, the bearing model generates proportional to its stiffness c_x in the displacement direction one spring force F_{Fx} and depending on the time derivative of the displacement $d\dot{x}$ proportional to a damping d_x a damper force F_{dx} in the displacement direction

$$F_x = F_{Fx} + F_{Dx} = c_x \cdot dx + d_x \cdot d\dot{x} \qquad (8.2)$$

which are then acting on both sides. This corresponds to the approach of a KELVIN-VOIGT-element discussed in more detail in Sect. 8.4.2.

Looking at the degrees of freedom of a chassis bearing, it can be seen that it can move in all three translational (axial, radial) and also in the three rotational (torsion, cardanic) directions. Movement naturally is not free, but is also limited by laws of force. Due to the bearing material and the geometric shape, a directional damping must be considered in addition to the directional stiffness.

In order to obtain these values, the bearing to be modeled must be measured. If this is done quasi-statically,[8] then progressive force-stroke characteristics are usually obtained for the stiffness curve. The free deformability of the material is usually restricted by metal sleeves, so that increasing deformation leads to an increase in stiffness, the progression.

---

[8]The term is, of course, formally incorrect. For the engineer, it means colloquially that the measurement of viscoelastic components is carried out so slowly that the damping components are negligibly low.

Fig. 8.14 One-dimensional
force law of a bushing

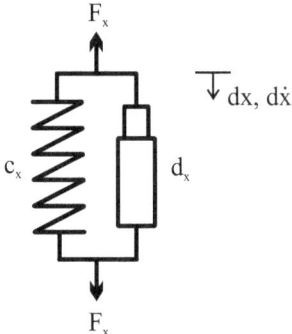

From the obtained data, either a linear or a non-linear parameterization can be carried out, depending on the respective question. How to do this and what the consequences are, will be discussed in the following two sections.

8.3.1 Linear Parameterization

The simplest approach to using bushings is to express each stiffness and damping parameter by a scalar value. However, this is only permissible if the deflections of the bearing are very small, so that the progressive stiffness curve is not used for larger bearing displacements. The workspace must therefore have a linear behavior so that it can be expressed by a scalar value. This is only the case if the scalar stiffness value was determined for a more or less fixed amplitude and frequency excitation and only this excitation, or in the sense of an operating point the direct environment of this excitation, is considered.

Bearings are generally used with preload, i.e. they are not force-free in the mounting position. Therefore, force or displacement offsets must be provided at least in the translational directions. In order to generalize the component model, some torque or angle offsets are also provided for the rotational degrees of freedom. Even for this approach, at least 18 parameters are generally required, as shown in Table 8.2.[9] If the bearing only has direction-independent radial rigidity and radial damping and the cardanic must also be described with one value each, the required number of parameters is reduced to 12 values.

The different stiffness in the radial direction and the cardanic are due to bearings which have different amounts of material in the two directions of force (kidney-shaped) or which have been made stiffer in one direction by vulcanized sheet metal strips (Fig. 8.15). Here it is absolutely necessary to compare the coordinate systems of the bearing in the model with

---

[9]In this example only a path or angle offset is specified. The force or torque offset is redundant. Depending on the measurement or modeling method, one or the other specification may be more advantageous.

Table 8.2 Linear
parameterization

Stiffness	Attenuation	Offset
c_{axial}	d_{axial}	$x_{0,axial}$
$c_{radial,x}$	$d_{radial,x}$	$x_{0,radial}$
$c_{radial,y}$	$d_{radial,y}$	$y_{0,radial}$
$c_{torsional}$	$d_{torsional}$	$\varphi_{0,torsional}$
$c_{cardanical,x}$	$d_{cardanical,x}$	$\Psi_{0,cardanical}$
$c_{cardanical,y}$	$d_{cardanical,y}$	$\theta_{0,cardanical}$

Fig. 8.15 Elastomer bearings
with different radial stiffness

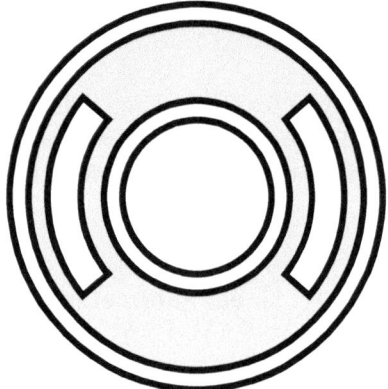

the design data and the measuring directions so that the stiffness match the true directions
of force in the mounting position.

At this point, it is not enough to know only the kinematic point of the bearing. The
orientation of the bearing also plays a role if the force directions have different stiffness.
You need a coordinate system that defines the axial direction and the two radial directions.
One convention could be that the z-direction of the coordinate system is always in the axial
direction and the radial x-direction in the link direction. The radial y-direction, in a right-
hand system, is then determined by the cross product of the z- and the x-direction.[10]

$$\vec{e}_z \times \vec{e}_x = \vec{e}_y \tag{8.3}$$

Depending on the design of the links, the radial x-direction is not necessarily the same as
the strut direction. For example, the axial direction of an A-arm can result from the
connecting line of the two bearings on the body side $(\overline{PQ})$, however, radial directions
still are uncertain. A third point is needed to determine the missing direction. In kinematics,
this is also known as the PQR-method, where P is the kinematics point, Q gives the axis of
rotation and R the orientation of the coordinate system on the rotation axis. Finally, these

---

[10]If the last active engagement with the cross product was some time ago: It is generally not
commutative, because the following applies $\vec{e}_x \times \vec{e}_z = -\vec{e}_y$.

Fig. 8.16 Determination of the orientation of the bearing using three points

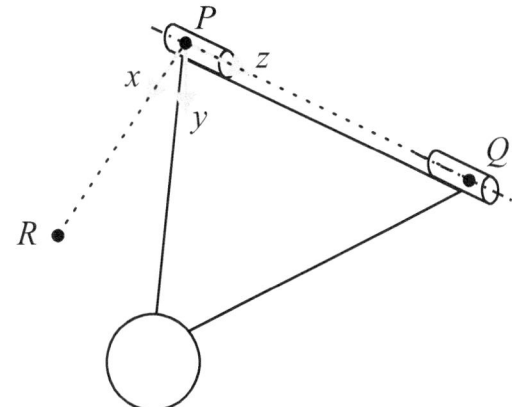

three points define a unique plane (Fig. 8.16). How the coordinate system of the bearing aligns itself to this plane can be interpreted differently in the different simulation environments and must therefore be checked at the beginning.

The stiffness of the bearing in any spatial direction can be determined by measurements. A force-stroke curve is then recorded during the measurement. Due to the amplitude and frequency dependence of the elastomers, several measurements should be performed covering the required parameter space with respect to the amplitudes and frequencies expected for the simulation. Thus, small amplitudes are required for linear vibration investigations (unbalance excitations) or for small road excitations (micro shake) and large amplitudes for bad road sections (load data determination).[11] Quasistatic, i.e. very low-frequency measurements facilitate the determination of static stiffness. Measurements in the natural axis frequency range cover the moving axis and measurements in the acoustic frequency range can be used to determine the transfer path of noise. So there is not one measurement for all load cases. Although obvious, this circumstance will frequently be *neglected*. If several calculation disciplines are interested in these measurements, the requirements must be combined in order to coordinate and reduce the number of required measurements as far as possible.

If one considers the measurements of the bearing, one must be aware that this is a viscoelastic component. Thus elastic (distance-dependent) and viscous (speed-dependent) components are always present simultaneously. Figure 8.17 shows this as an example for a bearing with constant stiffness and damping, as shown by the model approach from KELVIN-VOIGT (Sect. 8.4.2).

The viscous part is responsible for the hysteresis. Frequently, the average value from the two hysteresis branches is used for the stiffness curve. However, this is often only approximate.

---

[11]For small amplitudes, elastomeric bearings are usually harder than for large amplitudes. Sect. 8.3.3 explains how this can be handled.

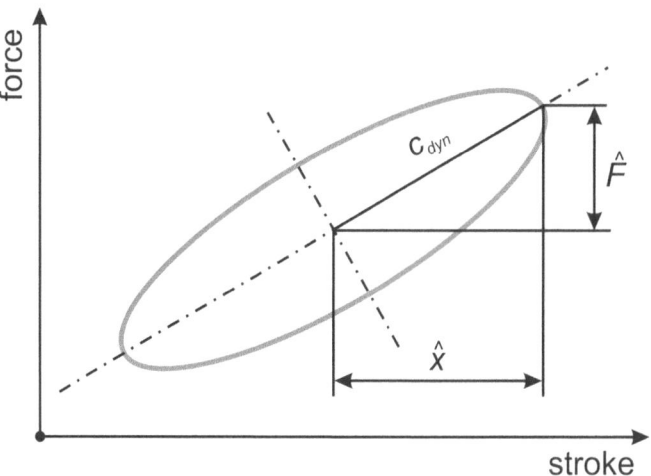

Fig. 8.17 Force-stroke characteristic of a linear elastomeric bearing

For a spring, the stiffness is simply determined from the quotient of the force amplitude and the stroke amplitude. In the case of the elastomeric bearing, the so-called dynamic stiffness c_{dyn} is thus obtained which has an elastic and a viscous part.

$$c_{dyn} = \frac{\widehat{F}}{\widehat{x}} \tag{8.4}$$

In the vibration theory,[12] the following approach is used to solve the underlying differential equation, which contains the previously unknown term δ:

$$c_{dyn} = \frac{\widehat{F}(t)}{\widehat{x}(t)} = \frac{\widehat{F}(\omega)}{\widehat{x}(\omega)} (\cos \delta + i \sin \delta) \tag{8.5}$$

The term loss angle δ is used to describe the damping of the elastomer bearing which illustrates the distortion between the deflection of the bearing and the delayed build-up of force (Fig. 8.18).

A loss angle of $\delta = 0°$ corresponds with an ideal elastic spring, an angle of $\delta = 90°$ an ideal viscous damper. The elastomeric bearings can therefore be found between these two extreme values. Equation (8.5) causes the dynamic stiffness to be calculated as the vector sum of the elastic component c and the viscous part $d\omega$.

---

[12]In vibration theory, the term complex stiffness is also used. The elastic part corresponds to the real part, the viscous part to the imaginary part of the equation.

Fig. 8.18 Loss angle

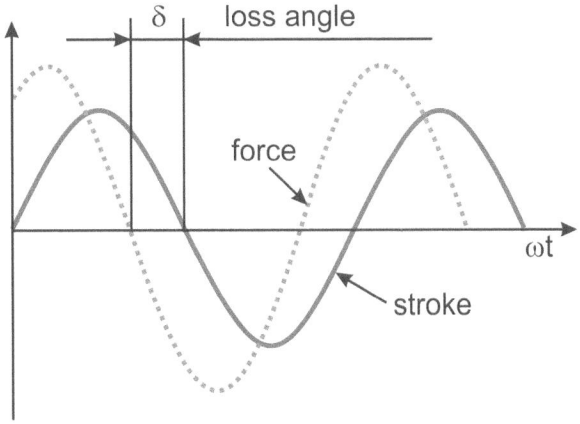

$$c_{dyn} = \sqrt{c^2 + (d\omega)^2} \tag{8.6}$$

For this reason, it is possible to approximately determine the static stiffness and damping of the bearing with known dynamic stiffness and known loss angle.

$$c = c_{dyn} \cos \delta \tag{8.7}$$

$$d = c_{dyn} \frac{\sin \delta}{\omega} = c_{dyn} \frac{\sin \delta}{2\pi f} \tag{8.8}$$

Equation (8.8) shows, on the one hand, that the dynamic stiffness depends on the excitation frequency f and that we still have an unknown here. The damping of a bearing can therefore only be linearized for a certain frequency.

In the case of a chassis bearing, the best measurement was made near the axle's natural frequency and this frequency value should then also be used.[13]

For elastomeric bearings, the loss angle is generally considered to be almost frequency-independent. An elastomer bearing thus dampens all frequencies equally well or no single frequency particularly.[14]

The loss angle thus has an influence on the dynamic rigidity of the bearing as shown in Eq. (8.6) and Fig. 8.19.

---

[13]The axle natural frequencies of normal cars are between 12 and 16 Hz. The value 15 Hz is often found in the literature. However, as far as known, this can of course be selected individually for the respective vehicle.

[14]For larger masses whose natural frequencies are known (engine, subframe), the hydromounts described in Sect. 8.5.1 are often used.

Fig. 8.19 Dynamic stiffness as a function of the loss angle

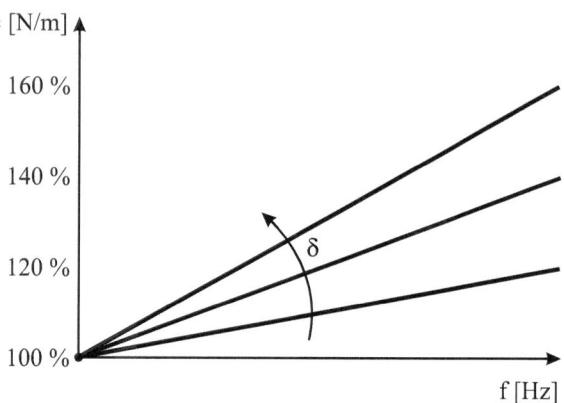

The material properties of elastomers can also be determined using the modulus of elasticity E which describes the relationship between the normal stress σ and the stretching ε.

$$E = \frac{\sigma}{\epsilon} \tag{8.9}$$

If we consider the phase delay from Eq. (8.5), this means for the stress using the addition theorem

$$\sigma = \sigma_0 \sin(\omega t + \delta) = \sigma_0 \cos\delta \sin(\omega t) + \sigma_0 \sin\delta \cos(\omega t) \tag{8.10}$$

This results in two components for the modulus of elasticity. One part reflects the elastic properties of the material. The so-called memory module E' is in phase to elongation.

$$E' = \frac{\sigma_0 \cos\delta}{\epsilon} \tag{8.11}$$

The second proportion gives the viscous proportion, which is $\pi/2$ shifted relative to the elongation and is also referred to as the loss modulus E''.

$$E'' = \frac{\sigma_0 \sin\delta}{\epsilon} \tag{8.12}$$

Both components are combined as a complex modulus of elasticity, from which the loss angle can also be read (Fig. 8.20).

The stiffness of the bearing is frequency and amplitude dependent, so that a linear representation can only represent one frequency and one associated amplitude at a time. If the bearing has been measured with different excitation frequencies, it is best to use the one that is closest to the axis natural frequency, as discussed above. If several excitation

Fig. 8.20 Modulus of elasticity and loss angle

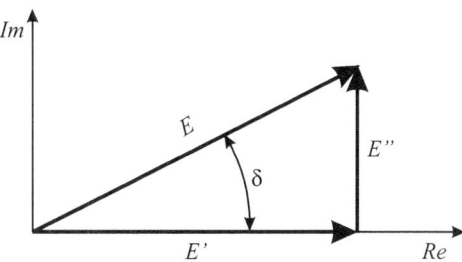

amplitudes were additionally measured, different stiffness can result at this frequency. In terms of ride comfort, this means that for low amplitudes, such as micro shake, other stiffness apply than for high amplitudes, such as on poor country roads or even on bad roads. A load-dependent parameterization should therefore be considered if the differences reach relevant orders of magnitude. If this is not desired, because there is additional potential for error,[15] more complex elastomer bearing models must be used.

8.3.2 Non-linear Parameterization

If you want to map the six degrees of freedom of the bearing completely using characteristic curves, then the above applies. Depending on the bearing type, 6–12 characteristic curves are required to describe the bearing completely (Table 8.3). However, the damping is usually not available as a force-velocity characteristic and the scalar value must also frequently be estimated using the loss angle via the phase delay, as described in the last section.

Most simulation programs allow you to choose whether the respective stiffness and damping are represented by a scalar value or a characteristic curve. However, it must be noted that some conversions permit both at the same time, so that the scalar and the characteristic curve values are superimposed.

However, this can also be used with a specific intention. If, for example, the progression of a stiffness is represented by a characteristic curve, the basic stiffness of the bearing can be represented by a linear value (Fig. 8.21).

If longer bearing travel is to be expected, as will be the case with most maneuvers of full vehicle simulations, at least the progressivity of the bearing stiffness must be taken into account. To do this, the bearing must be measured at a force level that extends into this progression.

It should be noted that even at very slow measuring speeds, slight hysteresis become visible in the force-stroke characteristics. It indicates that the damping properties of the

---

[15]If the load-case-dependent parameterization is changed manually, the probability of forgetting it is close to 100%. If the process can be automated, the possibility of errors is drastically reduced.

Table 8.3 Non-linear parameterization

Stiffness	Attenuation	Offset
$c_{axial}(x)$	$d_{axial}(xd)$	$x_{0,axial}$
$c_{radial,x}(x)$	$d_{radial,x}(xd)$	$x_{0,radial}$
$c_{radial,y}(x)$	$d_{radial,y}(xd)$	$y_{0,radial}$
$c_{torsional}(x)$	$d_{torsional}(xd)$	$\varphi_{0,torsional}$
$c_{cardanical,x}(x)$	$d_{cardanical,x}(xd)$	$\Psi_{0,cardanical}$
$c_{cardanical,y}(x)$	$d_{cardanical,y}(xd)$	$\theta_{0,cardanical}$

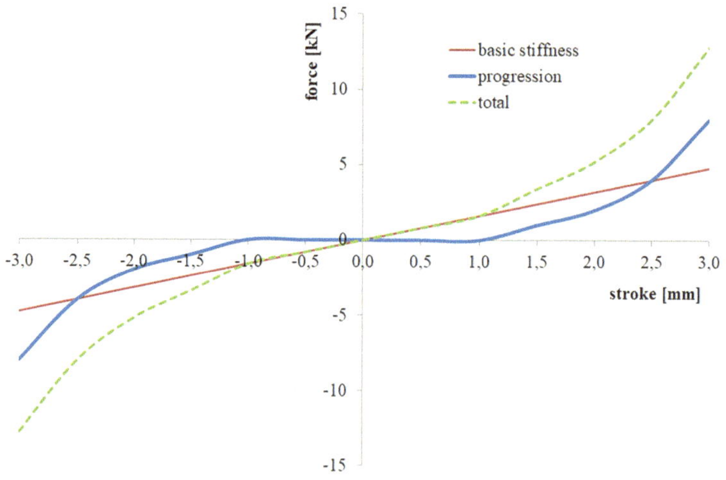

Fig. 8.21 Superposition of the basic stiffness and progression of an elastomeric bearing

bearing are already effective. In this context, what has been said in Sect. 6.1 shall be taken into account. A purely path-dependent force law cannot alone represent the mechanisms responsible for hysteresis (friction, viscous damping).

The already mentioned preload, with which chassis bearings are usually installed, can be taken into account by a path or angle offset or alternatively by an offset of the force or torque in the model. For parameterization with characteristic curves this means that either the characteristic curve passes through the zero point and is shifted with the offset[16] or that the offset is already taken into account in the position of the characteristic curve and accordingly the value of the offset must be set to zero.

The differences between the various options are shown in Fig. 8.22. As is so often the case, the measurement method and the modeling have to fit together in order to obtain the desired result.

---

[16]At this point at the latest, the sign convention of the simulation program used should be considered so that the bearing is preloaded in compression and not in tension.

Fig. 8.22 Indication of the force or travel offset of a characteristic curve

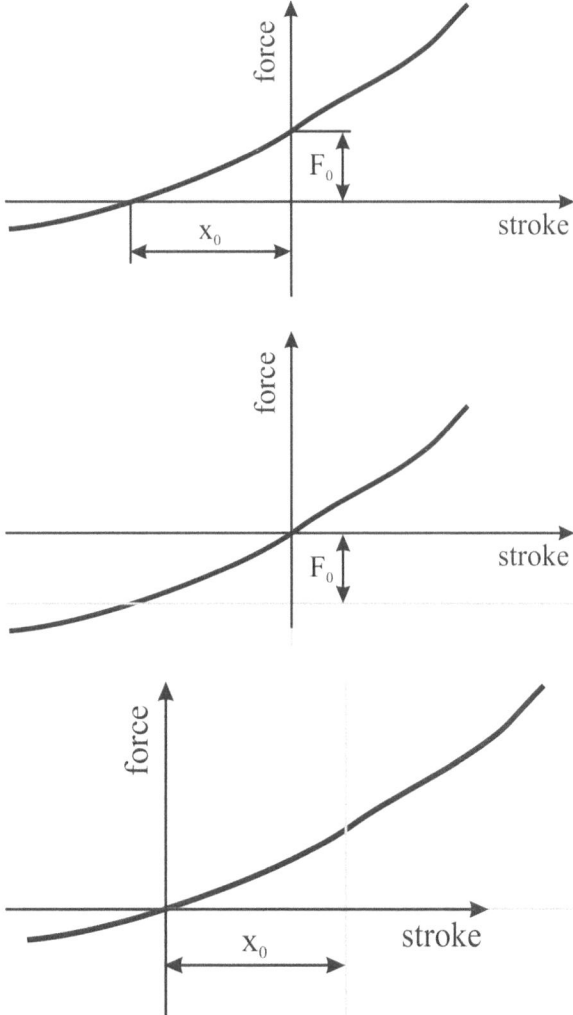

This is nothing else than the appeal not to blindly incorporate measurement data into one's own model, but to always be aware of how the measurements took place and what one's own model looks like.[17] The general procedure for creating and handling characteristic curves has already been dealt with in Sect. 6.1.5.

---

[17]With or without offset, that is the question . . .

8.3.3 Influence of Amplitude and Frequency of Excitation

If the same bearing is dynamically loaded with alternating excitation amplitudes and frequencies, a larger deviation of the stiffness curve compared to the static measurement is usually observed. When measuring, the same basic conditions must always be observed, as the elastomers behave significantly differently depending on their temperature and their previous history. The first load cycles will measure different stiffness than the later ones. Materials scientists call this behavior MULLINS- or PAYNE-effect, depending on whether large or small amplitudes are measured. In [Krei13] these effects and their causes are explained very clearly. Depending on the test rigs used, the measurement methodology determines the relationships between the maximum force amplitudes (necessary to achieve the progressions), the maximum displacement amplitudes (the test should be non-destructive) and the attainable frequencies.

Due to their amplitude and frequency dependence, the materials of the elastomer bearings thus exhibit strong non-linear behavior. These aspects are not covered by the classical path-dependent force laws. If one wants to use these models anyway (computing speed, costs, availability, etc.), then the use of a scaling factor offers a possible way out. Here a factor is determined which represents a scaling of the dynamic stiffness at the natural frequency of the system (e.g. in the axle range at approx. 15 Hz) with respect to the statically determined stiffness. Depending on material and amplitude, these factors typically vary between 1.0 and 2.0. A value of 1.0 can mean that the material, which hardens in small or medium path amplitudes (e.g. factor 1.4), then softens again in very large amplitudes and falls back into static stiffness. This factor can then be varied depending on the load case. In a quasi-static load case (e.g. test bench measurement) it is set to 1.0, in dynamic load cases (road crossing, amplitude-dependent) it may be increased to 2.0. This enables uniform parameterization of the bearing, which must then be varied depending on the load case. Since this behavior usually applies to several bearings in the chassis, this could be done via a central scaling factor. This approach is useful when averaged values over the observation period are of interest. This can be the case, for example, with evaluations in the frequency domain when crossing a stochastic road profile. If instead force curves are to be examined in detail directly at the bearing, then this approach will be too coarse. Especially if the excitation varies strongly between very small and very large amplitudes or very different frequency contents, larger differences to measurements can occur in the time domain.

For the investigations of compliance and for many load cases of vehicle dynamics simulation, the quasi-statically determined bearing characteristics are sufficient and the damping effect of the elastomers can often be neglected. Where the local vibration and force transmission behavior plays a greater role, i.e. in the ride comfort simulation or also in the load determination for the durability, the dynamic behavior of the bearing (in the relevant frequency range) must also be taken into account. For those whose focus is more on the bearing component or the introduction of force into the body or chassis, the frequency and amplitude dependence should be mapped in the model.

Further influences such as temperature or ageing of the material are also reflected in the stiffness characteristics and lead to a further increase in complexity. As with the other models, the extent to which and how detailed these dependencies must be represented in the bearing model depends strongly on the task of the required prediction accuracy. The simpler variant is to use a characteristic curve which represents the desired operating state and to compare it with the nominal state.

In order to be able to better map the material properties of the elastomer material, there are various approaches in rheology, some of which can also be found in the implementation of bearing models [Sedl01]. Their parameters must then be obtained by identification from measurements. The best known approaches are described briefly below. They are often found in parallel circuits and in combinations. The rheological basics necessary for this are very well described in [Mezg06].

8.4   Basics of Typical Elastomer Bearing Models

8.4.1   MAXWELL Element

The MAXWELL-element is used in rheology to represent a viscoelastic fluid by connecting a damper (NEWTON-element) and a spring (HOOKE-element) in series (Fig. 8.23). An external force acts simultaneously on the spring and the damper and both elements can be deflected independently of each other.

$$F = c\Delta s = d\Delta \dot{s} \tag{8.13}$$

The spring reacts immediately with a deflection, while the damper first builds up a movement through the acting force, which in turn can partially relieve the spring.

The change of path Δs of the system is thus composed of a part of the spring and a part of the damper.

$$\Delta s = \frac{F}{c} + \int \frac{F}{d} dt \tag{8.14}$$

If the external load is reduced, the elastic deformation of the spring is completely reduced. Residual deformation remains in the damper due to the viscous portion.[18] If the described behavior is ideally implemented, the system will react immediately with a

---

[18]The damper element used here should not be confused with the component of the shock absorber, which can also deform again due to the gas force acting on it. Here only the viscous part is included, which needs a speed to build up a force (or vice versa).

Fig. 8.23 Maxwell element

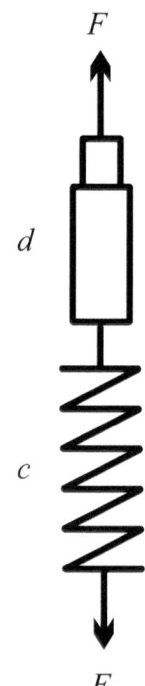

deflection of the spring when a force is applied (Fig. 8.24, left) and then continue to deform over time due to the damper (Fig. 8.24, right). This behavior is called creeping.

Real materials will show a time response, so the ideal abrupt change will be more like the dotted course. Accordingly, the system equations must be supplemented by exponential proportions as discussed in Sect. 4.5.

8.4.2 KELVIN-VOIGT Element

To model a viscoelastic solid by parallel connection of a viscous damper and a spring, the KELVIN-VOIGT-element can be used (Fig. 8.25). An external force acts immediately on both force members so that they can only move together due to the coupling.

$$\Delta s = \frac{F_F}{c} = \int \frac{F_D}{d} \, dt \qquad (8.15)$$

A delayed deflection occurs because the damper decelerates the movement of the spring. After the relief, the deformation is completely reduced, but again decelerated by the damping. The force to be applied is therefore the sum of the two parts.

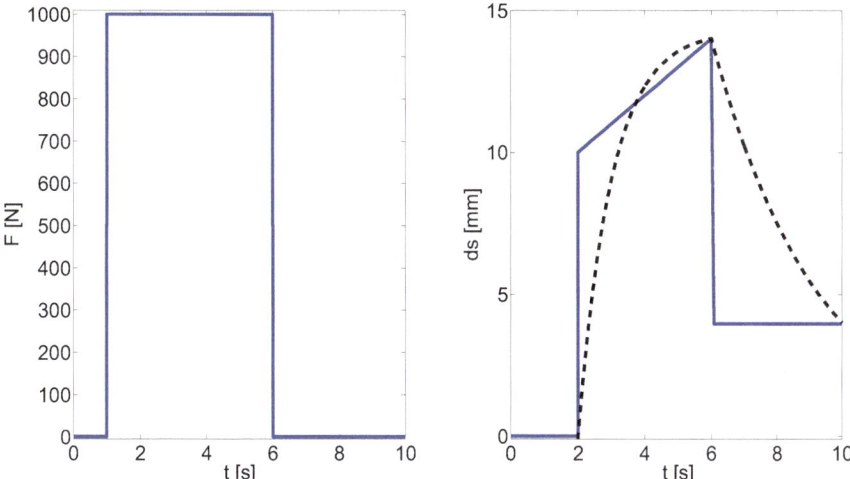

Fig. 8.24 Creep process

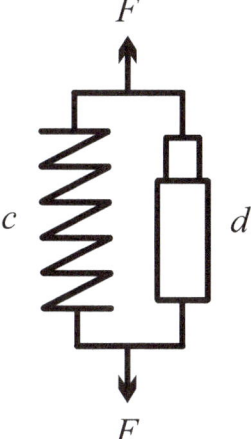

Fig. 8.25 Kelvin-Voigt element

$$F = c\Delta s + d\Delta \dot{s} \tag{8.16}$$

A rubber bump stop or elastomeric bearing can be regarded as a viscoelastic solid. In addition to their spring action, both components are mainly used due to the material damping.

8.4.3   Combination of Several Elements

The two basic models described can also be used in a network of several models in order to increase the accuracy of the model, but also the parameterization effort. More detailed investigations and model approaches can be found, for example, in [Sedl01].

8.5   Special Chassis Bearings

8.5.1   Hydromounts

In the chassis, hydromounts are mainly used where specific vibrations of certain frequency ranges are to be damped. Hydromounts are elastomeric bearings that also have a liquid volume that is pumped back and forth between two chambers. This results in significantly higher damping compared to pure elastomer mounts. In addition, this damping can be tuned very specifically to certain frequencies. This is always used if you know the natural frequency of the vibration system and want to damp it specifically (example: sub frame with axle or engine mounting).

In addition to the properties of the elastomeric bearings, a clear dependence on the excitation frequency can also be observed for hydromounts. Since hydromounts are specifically tuned to special frequencies, the damping effect is also greatest there (Fig. 8.26). In the case of rubber bearings, the change in the loss angle over the frequency is significantly less (Fig. 8.27). It should be noted here that only the relevant frequency range that can still be covered by MBS simulations is considered. Effects on acoustics are reserved for other disciplines.

In [NeLa99], the basic design of a hydromount is described using the example of an engine mount (Fig. 8.28). Special emphasis is placed here on the effect of the inflatable spring with particularly small and particularly large amplitudes by assigning a non-linear stiffness curve to it. The force to be transmitted consists of the proportion of the elastomer spring (KELVIN-VOIGT-model) and the proportion of the inflation spring due to the fluid effect.

$$F = c_x x_E + d_x \dot{x}_E + F_B \tag{8.17}$$

The reaction force of the inflation spring is determined from the following equation of motion.

$$F_B = A_T \, l_T \, \rho \, \kappa^2 \, \ddot{x}_P + d_T \, \kappa \, \dot{x}_P + d_{2T} \, \kappa^2 \, \dot{x}_P \, |\dot{x}_P| \tag{8.18}$$

with A_T: cross-sectional area of the ring channel between the chambers

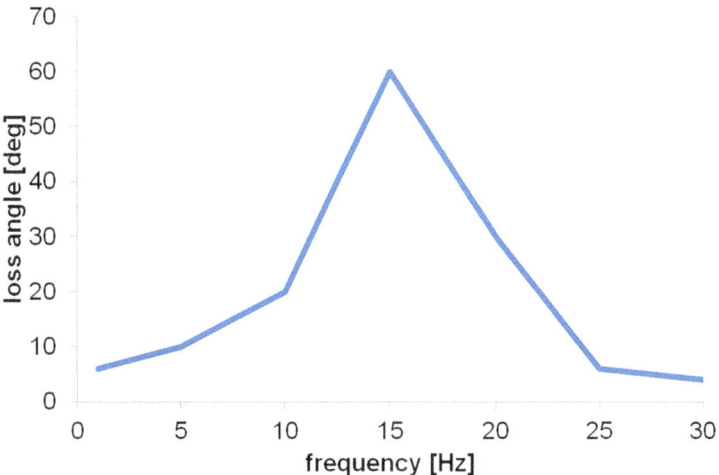

Fig. 8.26 Loss angle with hydromounts

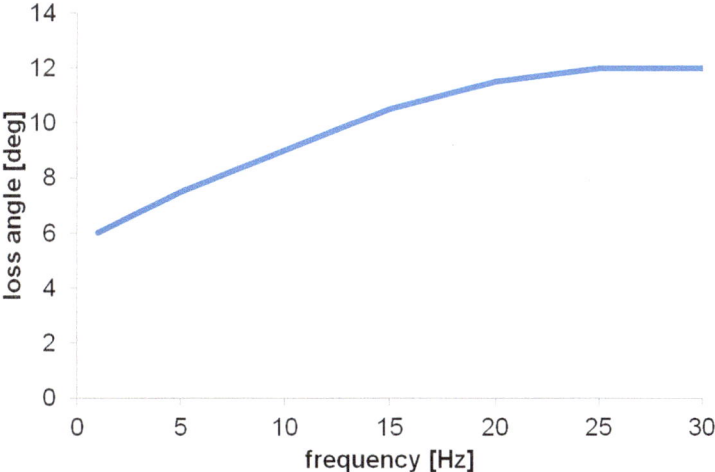

Fig. 8.27 Loss angle with rubber bearings

l_T: effective length of the ring channel

ρ : oil density

κ : ratio of the cross-sectional areas of the ring channel and the piston

d_T: viscous damping

d_{2T}: square viscous damping

The quadratic part of the viscous damping has its cause in the dissipation of the energy in the volume flow. If it is multiplied by the square of the velocity, it would always be

Fig. 8.28 Structure of a
hydromount according to
[NeLa99]

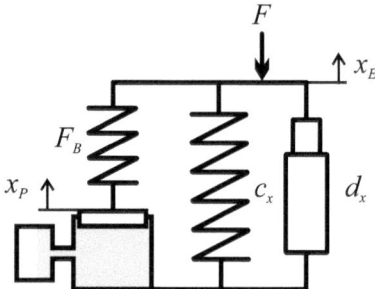

positive regardless of the velocity direction—which contradicts physics.[19] By multiplying
by the magnitude of the velocity (or alternatively by using the signum function), the
directional dependency can be taken into account [Dixo07].

Another model approach is described in [Zell12]. There, in addition, a loose part is
introduced, so that only the suspension spring can be used for small excitations. c_T
(elastomer bearing portion) and only with larger amplitudes is the proportion of the fluid
c_F being added (Fig. 8.29).

An approach for a semi-active hydromount, which is also used for an engine mount, can
be found in [HeGP07].

8.5.2 Top Mounts

The top mount represents the connection between the body and the main force path to the
wheel. At this point, the forces of the suspension spring, the suspension damper and any
additional springs, such as the bump or rebound stop, are applied on the vehicle body.

In principle, it can be regarded as an ordinary elastomeric bearing. Only the top mount
type must be taken into account. As shown in Fig. 8.30, force can be applied into the body
in one, two or three paths. For each of the up to three paths, the corresponding preloads
must be taken into account. In the zero position, the bump stop will normally be force-free,
in the damper the force of the compressed gas volume acts, and the suspension spring
supports the static wheel load.

Different stiffness and damping are used for each path. Ultimately, the top mount is
represented by one, two or three parallel elastomeric bearings which connect the body to
the respective force path. For deflection, the correct stop characteristics must be observed,
otherwise the bump stop cannot be brought to the correct force level.

---

[19]As you already learn in the first semester, friction is always opposite to the direction of travel.

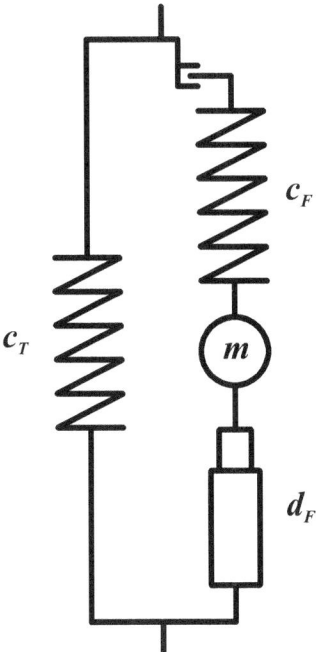

Fig. 8.29 Hydromount model according to [Zell12]

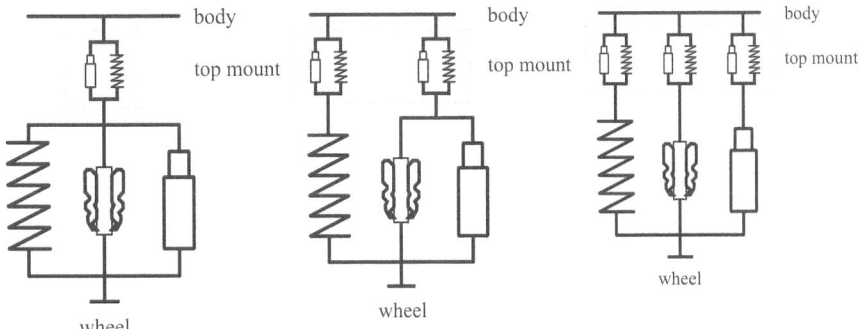

Fig. 8.30 Top mount types: one, two and three paths

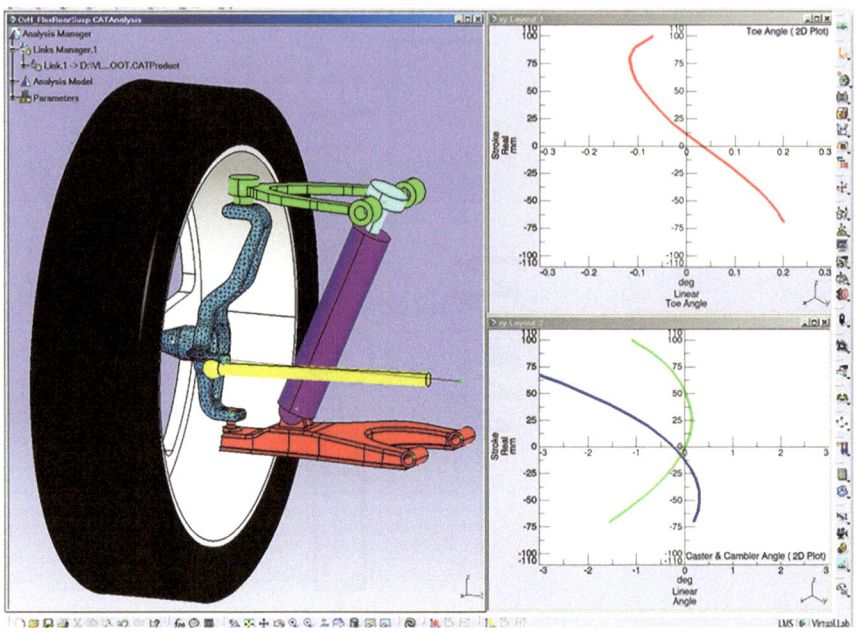

[Siemens PLM Software]

Fig. 8.31 Calculated wheel travel curves of a double wishbone suspension

8.6   Adjustment of Kinematics and Compliance with Measurements

8.6.1   Creation of Wheel Travel Curves

Since the kinematics and compliance of a suspension provide fundamental information about the driving behavior of the entire vehicle, a large number of standard measurements have become established in this area.[20] Some simulation environments that have been specially designed for vehicle simulation or that offer corresponding libraries already provide ready-made maneuvers and evaluation routines (Fig. 8.31). If this is the case, it must be clarified at the beginning whether the predefined procedures correspond to those in the real test. Great importance should be attached to this, because otherwise a direct comparison of a measurement and a calculation result is not possible satisfactorily. Of course, one can be converted into the other, but a discussion between test rig users and computational engineers will always be difficult. The more similar the evaluations are, the greater will be the acceptance of the results on the hardware side. Moreover, third parties

---

[20]Of course it is more the interaction between the front and the rear axle, but the measurements are usually done axle by axle.

(e.g. bosses) will rightly not understand why they are presented with the same facts in two different ways.

8.6.2   Deviations in Vehicle Level

The vehicle level plays a major role in both the kinematics and compliance tests. When comparing calculations and measurements, you will be able to see different levels immediately as a displacement of the resulting curves. Of course, one can shift one of the two curves computationally afterwards, but obviously, it is more favorable if you work with the same vehicle levels right from the start. The vehicle level can be adjusted via corresponding force or displacement offsets in the suspension spring.

8.6.3   Deviations in Kinematics or Compliance

If the toe or camber angle curves deviate from the measurement already in the zero position, the kinematics must first be checked. A three-dimensional coordinate measurement of the measured axle will usually not be available, so that on the model side one must make sure that the nominal coordinates of the link connections and the link lengths have been taken over correctly. Within the scope of the component and assembly tolerances, deviations from the nominal course may be justified. Therefore, depending on the type of deviation, it is recommended to slightly vary individual bearing coordinates. If the results approach each other, this can be an indication of the influence of the tolerances.[21]

Another possible cause is different preloads in the bearings, as this causes the force application points to shift. Here again it must be checked whether all nominal preloads were used in the model and if so, what influence the variations of these preloads have on the position of the toe and camber angle curve. In the case of bearings with direction-dependent radial stiffness, it must be clarified whether the mounting direction in the vehicle and model corresponds.

8.6.4   Additional Springs

The compliance measurements determine, among other things, the wheel-related stiffness of a wheel suspension. If one considers the force-stroke curve of an equilateral

---

[21]The variations must of course be within the actual component and assembly tolerances. The orders of magnitude should be clarified in advance with an expert.

measurement,[22] it is affected by hysteresis due to the friction and the damping effect of the elastomer bearings. The application points of bound and possible rebound bump stops can usually be estimated poorly. In this case, it is more advantageous to consider the derivative of this curve, since the stiffness is plotted over the path. Changes in stiffness, because this is what the points of application of the additional springs represent, can be better localized by steps in this curve (see also Sect. 9.3).[23]

While the tension springs are often linear auxiliary springs, which is shown in the stiffness-path diagram as a step to a higher constant level, the bump stops have a non-linear behavior. Each vehicle manufacturer has their own philosophy about when and how much to use the stops, but as described in Sect. 9.3, stiffness usually increases exponentially. If a component measurement is available, you can check it with this maneuver. If it is not available, the stiffness curve of the buffer can be iteratively adapted to the vehicle measurement. In this force path, however, lies also the top mount, which was dealt with in the last section. In addition to the correct connection of the force path and its path limitation, a comparison of the stiffness must show which effects belong to the additional springs and which to the top mount.

8.6.5 Suspension Spring Stiffness

In the zero or design position range, only the suspension springs act on most vehicles. The additional springs are therefore not yet engaged. For this reason, it is best to read at this point, even if only indirectly, whether the correct stiffness was selected for the suspension spring and whether the expected spring was installed in the measured vehicle. If the mainspring is a linear steel spring, the stiffness value around this position must be constant. In the case of a non-linear stiffness curve, a displacement may indicate a different vehicle level in the measurement and the calculation. The amount of stiffness can only be read off indirectly, since the so-called secondary suspension or parasitic suspension spring was also measured. If a measurement was carried out with the suspension spring removed, the proportion of the secondary suspension spring can be estimated.[24] More information on this can be found in Sect. 9.3.3.

If an air spring is used for the suspension spring, it will behave approximately isothermally due to the slow travel speed, since the heat generated by the compression can be

---

[22]The left and right halves of the axle are sprung in and out at the same time with the same amplitude, so that any anti-roll bar that may have been installed does not work. If an anti-roll bar is installed and its bearings are vulcanized on, the torsional stiffness of the anti-roll bar bearings is part of the secondary suspension spring. Unless it is explicitly available, it can be mapped in the anti-roll bar bearing.

[23]These are wheel-related application points that must first be converted to the component.

[24]Since the elastomeric bearings are then no longer tensioned with the same preload, the stiffness of the secondary suspension spring in the zero position can only be approximately determined.

released sufficiently quickly from the air spring system into the environment. This means that the stiffness of the spring is lower than with a dynamic load. Since the stiffness of the air spring is also load-dependent and, due to the rolling contour, travel-dependent, it is even more important in this case to compare the vehicle level with the measurement. If no complex air spring model is used which takes all these aspects into account, the appropriate characteristic curve for this load case must be selected. More details on these effects can be found in Sect. 9.2.

8.6.6 Anti-roll Bar Stiffness

If a reciprocal measurement is considered, the stiffness of an anti-roll bar can be seen in the diagram in addition to the equilateral excitation, if installed. This is an excellent way to check the anti-roll bar stiffness used in the model or, if it is not available as a value, to determine it with this maneuver. See Sect. 9.1.4 for details on the anti-roll bar component.

References

[Dixo07]   Dixon, J. C.: *The Shock Absorber Handbook*, John Wiley & Sons Ltd, Chichester, 2007

[HaHo03]   Halfmann, C. and Holzmann, H.: *Adaptive Modelle für die Kraftfahrzeugdynamik*, Springer, Berlin, 2003

[HeGP07]   Heimann, B., Gerth, W. and Popp, K.: *Mechatronik Komponenten – Methoden – Beispiele*, Fachbuchverlag Leipzig in Carl Hanser Verlag, Munich, 2007

[Henk93]   Henker, E.: *Fahrwerktechnik. Grundlagen, Bauelemente, Auslegung*, Vieweg, Braunschweig, 1993

[Ise192]   Isermann, R.: *Identifikation dynamischer Systeme, Band 1; Grundlegende Methoden*, Springer, Berlin, 1992

[Krei13]   Kreiselmaier, R.: Mullins oder Payne – Zwei „starke" Effekte der Gummielastizität, in *FFD im Dialog*, 23 – 28, company publication Freudenberg Forschungsdienste, Weinheim, 2013

[Kroh01]   Krohmer, A.: Einfluss der globalen Karosserietorsionssteifigkeit auf die Fahrdynamik, in *Reifen-Fahrwerk-Fahrbahn*, VDI-Berichte 1632, Hannover, 347-371, 2001

[Kune05]   Kunert, A. et al: Echtzeitfähige Achsmodelle: Praxisbeispiele von der Konzeptstudie bis zur Validierung, in Reifen-*Fahrwerk-Fahrbahn*, VDI-Berichte 1912, Hannover, 85-94, 2005

[Lang97]   Lang, H.P.: *Kinematik-Kennfelder in der objektorientierten Mehrkörpermodellierung von Fahrzeugen mit Gelenkelastizitäten*, Fortschrittberichte VDI, Reihe 12, Nr. 323, Duesseldorf, 1997

[LiSe01]   Lion, A. and Sedlan, K.: Anwendungen spezieller Komponentenmodelle zur Simulation von Fahrwerkbelastungen auf Schlechtwegstrecken und Komfortanalyse, in *Reifen-Fahrwerk-Fahrbahn*, VDI-Berichte 1632, 225-253, 2001

[Mats07]   Matschinsky, W.: *Radführungen für Straßenfahrzeuge*, Springer Verlag, Berlin, 2007

[Mezg06]   Mezger, T.G.: *Das Rheologie Handbuch. Für Anwender von Rotations- und Oszillations-Rheometern*, Vincentz Network, Hannover, 2006

[NeLa99]   NEUREDER, U. AND LAERMANN, F.-J.; Minimierung des Motorstuckerns am Beispiel des
             Ford Fokus, in *Reifen-Fahrwerk-Fahrbahn*, VDI-Berichte 1494, Hannover, 253-267,
             1999
[ReBe00]   REIMPELL J. AND BETZLER J.W.: *Fahrwerktechnik: Grundlagen*, Vogel Verlag,
             Wuerzburg, 2000
[ScHB10]   SCHRAMM, D., HILLER, M. AND BARDINI, R.: *Modellbildung und Simulation der Dynamik
             von Kraftfahrzeugen*, Springer, Berlin, 2010
[Sedl01]   SEDLAN,   K.:   *Viskoelastisches   Materialverhalten   von   Elastomerwerkstoffen:
             Experimentelle Untersuchung und Modellbildung*, Dissertation, University of Kassel,
             2001
[Tobo04]   TOBOLÁŘ, J.: *Reduktion von Fahrzeugmodellen zur Echtzeitsimulation*, Dissertation,
             University of Prague, 2004
[Zell12]   ZELLER, P. (ED.): *Handbuch Fahrzeugakustik*, Vieweg+Teubner, Wiesbaden, 2012

Springs

Elastomer bearings (rubber-metal parts) can also be regarded as springs. After all, they make a significant contribution to secondary suspension in multi-link axles (Fig. 9.1). However, they have already been described in Sect. 8.3, so that in this chapter the springs in the vertical force path and the anti-roll bar step into the foreground. First, you have to think about which aspects of the spring you want to reproduce. Must the shape, the mass properties and/or the law of force be mapped? In the sense of multi-body systems, springs are pure force elements—without mass properties.

However, in order to take their mass into account, they can, for example, be added to one half of the vehicle body and to the other half of the wheel suspension.[1] This *trick* is only used, however, to ensure that the mass is reflected in the total sum of the vehicle. The position of the replacement mass close to the component has only an insignificant influence on the vehicle's center of gravity. Due to the division in half, for example, one part can be found in the sprung mass and one part in the unsprung mass. The correct distribution of the unsprung mass plays a role, especially when the road surface is uneven or wheel load fluctuations are considered. It would be fatal, for example, if the spring weighed 2 kg and its mass was completely added to the 2 kg link. The effective inertial forces would then clearly be different to the original.

---

[1]This may be difficult for the moments of inertia and the change of the center of gravity position, which is why they are often forgotten.

© Springer Fachmedien Wiesbaden GmbH, part of Springer Nature 2021
D. Adamski, *Simulation in Chassis Technology*,
https://doi.org/10.1007/978-3-658-30678-6_9

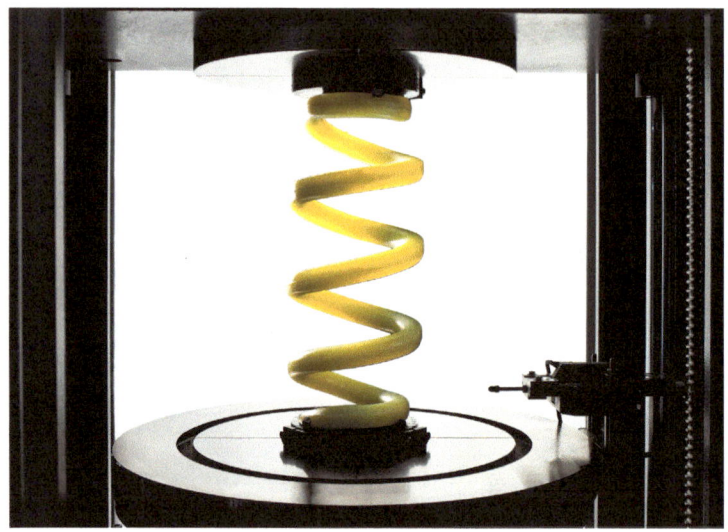

[Audi AG]

Fig. 9.1 GRP spring clamped in test bench

9.1   Steel Springs

9.1.1   Coil Spring

In most passenger cars, the suspension spring, which supports the body in relation to the chassis, is designed as a coil spring. In the following, the assumption is made that this spring is always under a preload and is prevented by its shape or by the bump stop from going on block. That way, effects such as coil strikes, lifting of the spring ends or collisions with other components can be excluded. These effects therefore do not have to be modeled.

9.1.1.1 Force Law
When modeling a coil spring, it must first be clarified whether it is a linear or a non-linear spring. The force law of the linear spring (Fig. 9.2) can be represented in good approximation by a constant of the spring stiffness (Eq. 9.1), in the non-linear variant (Fig. 9.3) the spring stiffness must be present as a force-stroke characteristic curve (Eq. 9.2).

$$F(ds) = c \cdot ds \tag{9.1}$$

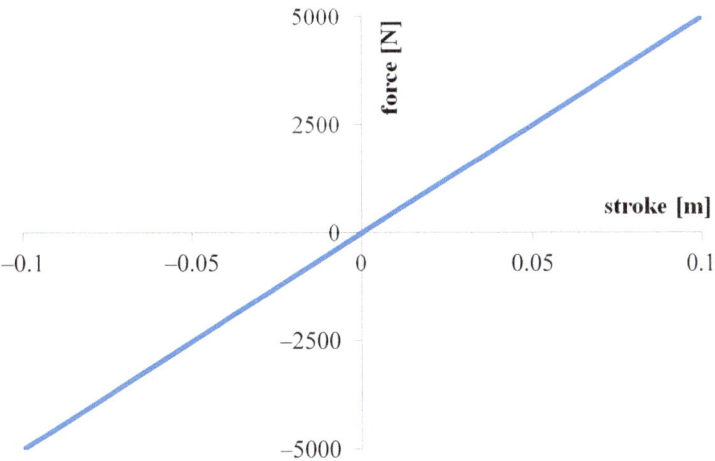

Fig. 9.2 Force law of the linear spring

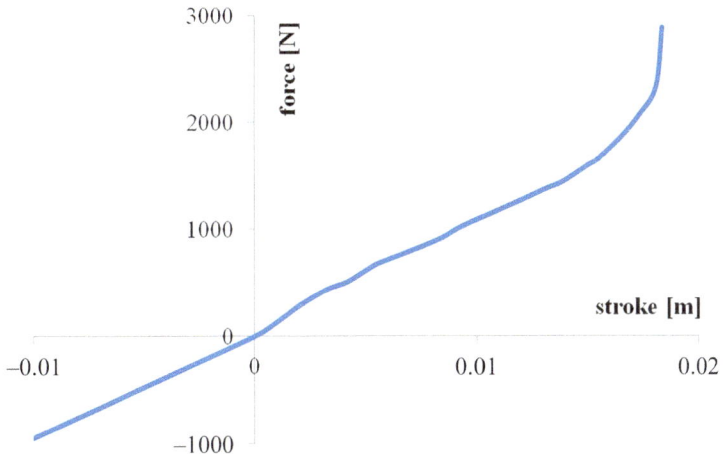

Fig. 9.3 Force law of the nonlinear spring

This simple representation is sufficient for most MBS-relevant load cases. This one-dimensional force law reacts statically and dynamically, with small and large amplitudes always the same—as long as one remains in the working range of the spring, no buckling or a block must be taken into account. The characteristic curve is easy to create.

$$F(ds) = c(ds) \cdot ds \tag{9.2}$$

However, the causes leading to the non-linearity are no longer considered in the following. If the non-linearity is caused by a variable wire cross section, this changes the vibration behavior of the spring over its spring deflection. However, this effect is usually negligible for a modeling of the suspension spring. Nevertheless, one must consider whether these effects are relevant or not for the question to be considered before modeling.

In the described one-dimensional modeling in the MBS simulation programs, the spring is regarded as massless and without kinematic influence. It cannot support longitudinal or transverse forces and therefore has no kinematic function. If the spring is interesting as a component and not only in its force effect, this no longer applies—more complex approaches can be found in most commercial simulation environments. They are mostly used in the simulation of valve springs, where the inertia of the spring mass plays a not insignificant role at the high excitation frequency. This also applies if the bending moment to be transmitted by the spring cannot be neglected. In [MeSc07] some examples of more complex models for these purposes are listed.

The standard modeling does not provide for any friction or material damping. The latter can certainly be neglected with a steel coil spring.

If the non-linearity of the spring is caused by additional springs (bound and rebound bump stops), it should be considered whether it is advisable to include their stiffness curve in the spring characteristic curve or to model them as separate components. The consequences of this are described in more detail in Sect. 9.3.

9.1.1.2 Graphical Representation

In commercial simulation environments, you often find graphical representations of coil springs, which are only used for graphical representation and have only a limited relation to the actual suspension behavior. Nevertheless, this representation is helpful for understanding the movement.

9.1.1.3 Penetration

Compared to the real spring, the MBS spring does not have the restriction that it must not go on block. Since it is only a law of force, it does not have corresponding contact models that would detect the contact of the windings. That way, the spring can penetrate itself and work with negative paths and thus negative forces according to Eq. (9.1). In the case of wheel suspensions, this is generally not possible due to the design. If the first attempts are made to model a single-mass oscillator, this effect has to be considered. One should, however, always be aware of this effect. As you can see in the next section, a spring length of zero may not be an unusual convention for the design position. That way, positive paths generate tensile forces and negative compressive forces. Since the displacement from zero would not provide any force, a constant force offset must represent the static load.

9.1.1.4 Free Spring Length

If the spring is installed in a concrete axle kinematic system, the free length must be taken into account. The springs are installed under a compressive force because they must already carry the static weight in the design position.

Depending on the simulation environment, the free length of the spring can be specified so that a force offset occurs due to the installation position of the spring. Alternatively, the spring preload (force offset), which the spring has in the mounting position, can often be specified directly. It is usually identical with the static wheel load (converted to the point of application of the spring force). The force-stroke curves in Figs. 9.2 and 9.3 pass through the zero point, which means that they represent the force F of the spring during a deflection ds which are represented by an offset F_0 has been corrected. The law of force extends, based on Eq. (9.1), with the free length s_0 too

$$F(ds) = c \cdot (ds - s_0) + F_0 \qquad (9.3)$$

The same applies to the non-linear spring. The sign must be taken into account, since if the spring is elongated ($ds > 0$) tensile forces during compression ($ds < 0$) compressive forces occur.

9.1.1.5 Isolation

If the spring is elastically isolated from the vehicle body and/or the suspension, this should be taken into account as an additional component or as a law of force if possible (see also Sect. 8.3). If the separate figure is not relevant for the problem or if there are no stiffness for this bearing arrangement, then it should be considered how the so-called secondary spring rates are to be taken into account in the overall model.

9.1.1.6 Complex Spring Models

In the approaches described so far, the spring is therefore not seen as a component but rather formulated as a law of force. If this is not sufficient because, for example, contact processes, coil shocks or the lifting of the spring ends must be taken into account, more complex spring models must be used, such as those described in [Witt06] or [MeSc07].

9.1.2 Leaf Spring

The main difference between the leaf spring and the coil spring is that it also has a kinematic function. It can transmit longitudinal and transverse forces and thus provide support for the wheel in these directions. In addition, it is attached on two body points, which must be taken into account in its connection. Depending on the design, more or less friction occurs between the leaves, which must also be modeled if the response behavior at small amplitudes is to be taken into account. This friction is sometimes used explicitly as a

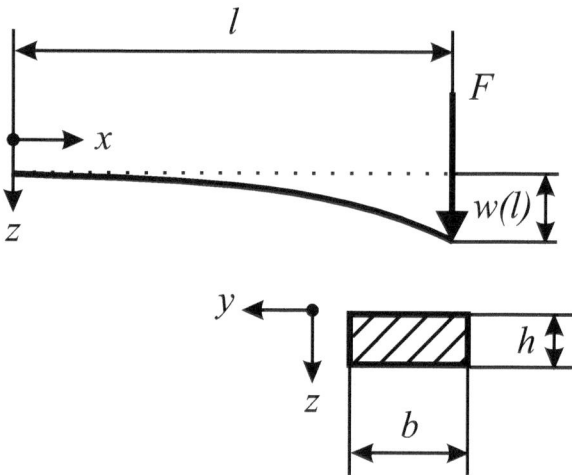

Fig. 9.4 Bending line of the beam clamped on one side

damping element, so that both the static and the dynamic friction behavior must be taken into account.

9.1.2.1 Force Law

The simplest way to represent a single leaf spring is to assume that only a vertical force is applied. F acts on a beam clamped on one side. The calculation of the corresponding bending line is material in the lecture of strength theory. The bar shown in Fig. 9.4 represents half of a single leaf of a rectangular leaf spring.

The stiffness is calculated using the differential equation of the bending line.

$$w''(x) = -\frac{M(x)}{E\,I(x)} \tag{9.4}$$

With the assumed rectangular cross-section, the moment of inertia of the surface I is constant over the entire length.

$$I(x) = I = \frac{b\,h^3}{12} \tag{9.5}$$

Through twofold integration and consideration of the boundary conditions, one obtains the following equation for the deflection w at the end of the spring

$$w(l) = \frac{F\,l^3}{3\,E\,I} \tag{9.6}$$

If the deflection and the effective force are now put into proportion, the stiffness of the blade is obtained for a rectangular cross-section.

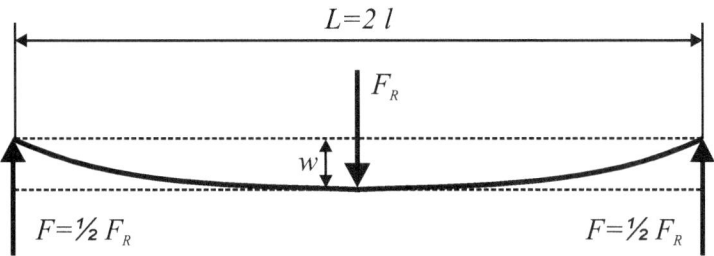

Fig. 9.5 Bending line of the leaf spring clamped on both sides

$$c = \frac{F}{w(l)} = \frac{3\,E\,I}{l^3} = \frac{b\,h^3\,E}{4\,l^3} \tag{9.7}$$

For the entire spring (Fig. 9.5), the following applies to the deflection with a symmetrically applied wheel load F_R

$$w(l) = \frac{F_R\,L^3}{48\,E\,I} = \frac{F_R\,l^3}{6\,E\,I} \tag{9.8}$$

Even though the rectangular section is easy to calculate, it is not used due to its high weight. The more common designs are trapezoidal and parabolic leaf springs. The significantly better ratios of weight and stiffness.

With the trapezoidal leaf spring, the leaves become narrower towards the end at a constant height (Fig. 9.6, central). In the limit case of the triangular spring, the final width is b_1 to zero, which of course is not feasible in this form. In contrast to the rectangular spring, the moment of inertia of the surface changes over the x-coordinate, so that with the initial width b_0 according to [MeSc07] applies

$$I(x) = \frac{b(x)\,h^3}{12} = \frac{h^3}{12}\left[b_1 + \frac{x}{l}(b_0 - b_1)\right] \tag{9.9}$$

For the calculation of the deflection at the spring end the width ratio $\beta = b_1/b_0$ was introduced.

$$w(l) = \left[\frac{1 - 4\beta + 3\beta^2 \ln\beta}{(1 - \beta)^2}\right]\frac{6\,F\,l^3}{E\,b_0\,h^3} \tag{9.10}$$

The stiffness of the trapezoidal leaf spring then results from the relationship between the applied force F and the deflection at the end $w(l)$.

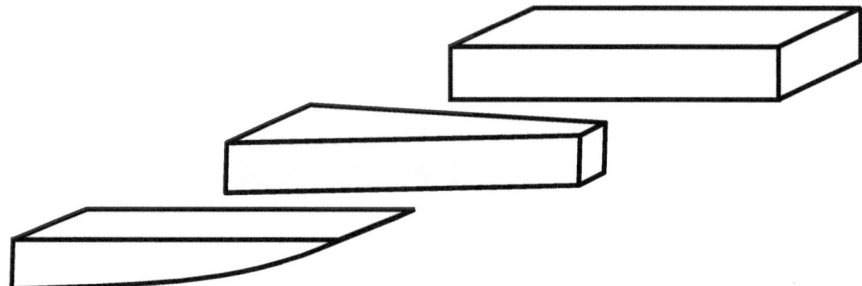

Fig. 9.6 Leaf spring designs (parabolic, trapezoidal, and rectangular)

$$c = \frac{F}{w(l)} = \left[\frac{(1-\beta)^2}{1 - 4\beta + 3\beta^2 \ln \beta}\right] \frac{E \, b_0 \, h^3}{6 \, l^3} \tag{9.11}$$

With the parabolic leaf spring, the height is reduced parabolically towards the end while the width remains the same (Fig. 9.6, left).

$$h(x) = h_0 \sqrt{1 - \frac{x}{l}} \tag{9.12}$$

According to the above procedure, the parabolic spring has the following stiffness:

$$c = \frac{F}{w(l)} = \frac{E \, b \, h_0^3}{8 \, l^3} \tag{9.13}$$

If the contour changes cubically

$$h(x) = h_0 \sqrt[3]{1 - \frac{x}{l}} \tag{9.14}$$

the spring becomes stiffer with the same dimensions.

$$c = \frac{F}{w(l)} = \frac{E \, b \, h_0^3}{6 \, l^3} \tag{9.15}$$

The previous calculations were carried out for individual sheets. In order to achieve non-linear behavior and/or greater stiffness, the leaves are layered. When the spring moves, the individual layers move relative to each other and friction forces arise. The resulting dissipation is and has been often referred to as a *damping* which partly saved additional dampers. The permanent transition from static to dynamic friction, however, does not provide a good response or satisfactory comfort, so that today plastic inserts are often

Fig. 9.7 Layered leaf spring

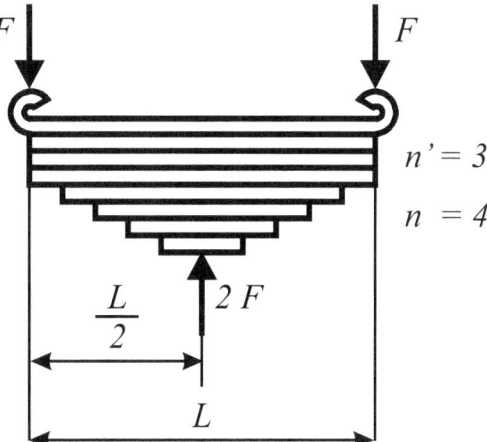

placed between the leaves in order to minimize the friction effects. Nevertheless, in the case of multi-layer leaf springs, frictional forces should be measured in order to estimate whether these effects need to be modeled. A brief comment on this subject follows in the next section.

In [MeSc07], the following relationship has been established for a layered rectangular leaf spring with symmetrical load for the spring deflection

$$w = \frac{12\, l^3}{(2n + n')\, E\, b\, h^3} \tag{9.16}$$

where n the number of layers and n' the number of additional layers of full length $L = 2\, l$ (Fig. 9.7).

Meanwhile, there are some developments to manufacture the leaf springs from fibre composite materials, in order to get primarily the weight problem but also the friction under control. If such a spring is the subject of consideration, the formulas given above must be adapted to the respective design. Steel and iron are easier to describe. In [Fruh10], such an approach of a wheel-guiding transverse leaf spring is presented.

The law of force can also be represented by special elastic beam attachments, which, depending on the software, are massless or mass-afflicted. The numerical stability, computational time requirements and parameter requirements of these models must be critically examined beforehand to see whether they provide added value compared with the *simple* laws of force.

Of course, the mass of a leaf spring can no longer be neglected, especially in commercial vehicle construction, which means that it may have to be closed on the vehicle body and/or axle side.

9.1.2.2 Friction

It becomes more problematic with the friction. Frequently, a damping parameter or a damping matrix can be specified in addition to the stiffness-determining variables. However, this corresponds to viscous damping, i.e. speed-dependent damping, and has nothing to do with the effects of static and dynamic friction. If no additional friction models are to be integrated, linearization by one operating point is often used. A constant force offset is determined from a hysteresis measurement, which is then added to the spring force. It should be noted that the directional dependency is maintained[2] and that one can only assume a load case in which the spring is always broken loose.[3] This is the case, for example, when driving on bad roads. A more detailed description of friction can be found in Sect. 10.2.

A detailed study on friction and hysteresis in the context of static and dynamic stiffness of leaf springs is described in [Lee03].

9.1.2.3 Kinematics

The kinematic function of the longitudinal and transverse support can be improved by combining the law of force with joints or explicit constraints. The deformation of the spring in the longitudinal and transverse directions has an influence on the wheel guidance, which can lead to larger errors, especially with braking and larger transverse forces, if this is not taken into account.

9.1.3 Torsion Bar

The modeling of a torsion bar can be done analogous to the coil spring, but instead of a force-stroke characteristic, a torque-angle characteristic must be specified. If the stiffness is to be determined from the material values and the dimensions, then with pure torsion loading and a circular cross section with the diameter of d and the length l for a given shear modulus G

$$c_\varphi = \frac{\pi \, G \, d^4}{32 \, l} \tag{9.17}$$

---

[2]For example by a signum function, which can be numerically problematic.
[3]This means that the spring forces are significantly higher than the breakaway force from the static friction.

9.1.4   Anti-roll Bar

The anti-roll bar is ultimately a torsion bar with the special feature of being bound to the deflection of the two wheel suspensions of an axle. With alternating suspension, this results in twisting, which then leads to a torque which is retained by the vehicle body in order to reduce its roll angle. During implementation, care must be taken to ensure that the anti-roll bar model can cover various scenarios.

The first thing that always comes to mind is the main application. When cornering, the centrifugal force acting on the center of gravity causes

$$F_C = m \cdot \frac{v^2}{r} \qquad (9.18)$$

and with the distance between the center of gravity and the roll axis a torque occurs that rotates the body about the roll axis. This torque is countered by the spring effect of the body suspension and the anti-roll bar. If the effect of the anti-roll bar is reduced to a torsional spring stiffness that is proportional to the torsion, the actual principle of action is neglected.

It is essential for the anti-roll bar that there is uneven spring deflection on the right and left sides of the same vehicle axle. With the same deflection, it does not act accordingly. This could still be achieved with a twisting angle of zero, but there is another undesirable scenario. If the vehicle drives straight ahead, but on an uneven road surface, the two wheels of one axle also spring in unevenly. Again, a torque arises that this time affects the structure in such a way that it will rotate around the roll axis. This undesirable effect is also called *throwing* the body. It is therefore imperative that the deflection of the wheel is also taken into account when driving over uneven road surfaces.

9.1.4.1 Force Law

The deformation of the anti-roll bar takes place during the deflection process through a combination of bending and torsion moments. Due to the frequently multiple bent contour (Fig. 9.8), a simple analytical determination of the anti-roll bar stiffness is usually not possible.

In the simplest modeling, a linear relationship is established between the reciprocal deflection $\Delta x = x_l - x_r$ and the generated force or between $\Delta \alpha = \alpha_l - \alpha_r$ and the generated torque (Fig. 9.9). The linear equivalent stiffness well reproduces the behavior around the resting position. For most, rather small deformations, this approach is sufficient. In the case of larger spring deflections, however, deviations from the real component occur increasingly, so that it is necessary here to rely on component measurements or on the results of non-linear FEA calculations.

For the anti-roll bar there is also the possibility to use complex bar models to better map the contour. Here, too, however, the stability, computing time and parameter requirements must be critically examined beforehand to determine whether the use of the complex model provides added value compared with simple force laws.

Fig. 9.8 Curved contour of the anti-roll bar on the rear axle of a Mercedes S-Class (W221)

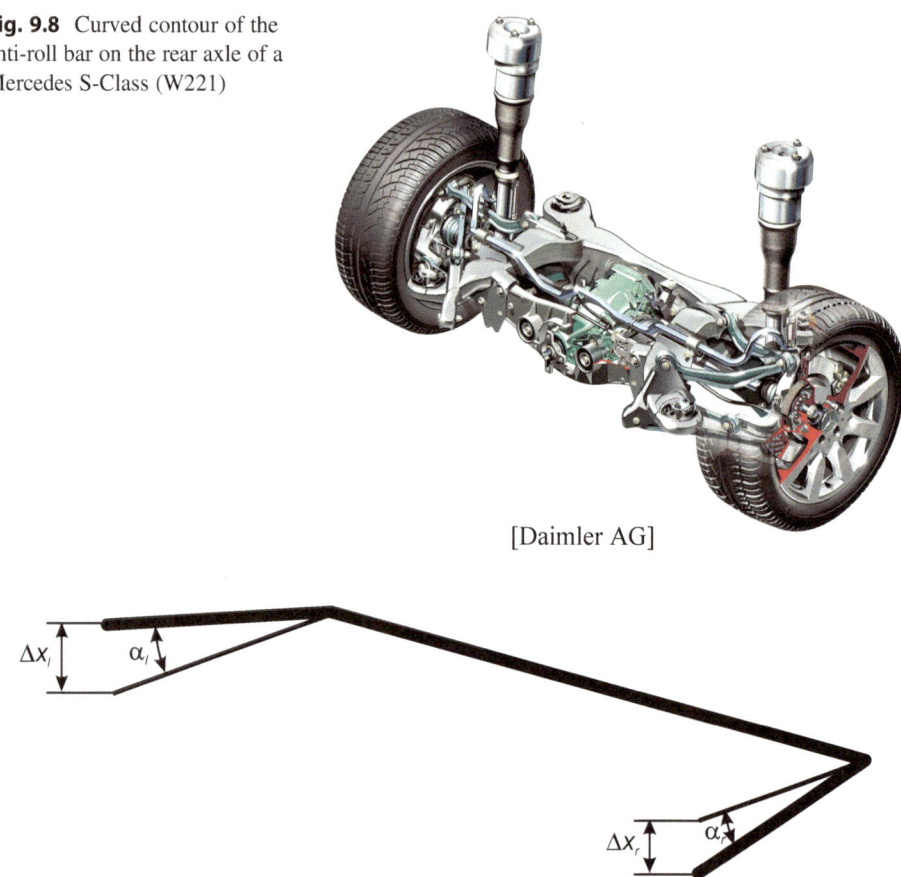

[Daimler AG]

Fig. 9.9 Alternate deflection of the anti-roll bar

9.1.4.2 Anti-roll Bar Mounting

The anti-roll bar is usually mounted to the sub frame or vehicle body with elastic bearings. A distinction must be made between whether the anti-roll bar can rotate in the bearing (only radial forces are transmitted) or whether the bearing has been vulcanized (axial forces and torques can also be transmitted).

If the model of the anti-roll bar is to be parameterized on a K&C test bench with the aid of a full vehicle measurement, it must be taken into account that the forces and torques resulting from the mounting were also measured. If individual measurements of the anti-roll bar and the bearings are available, parameterization is correspondingly simpler.

9.2 Air Spring

The depth of the modeling of an air spring depends strongly on the simulation task and the air spring to be modeled. Is there an additional volume that can be switched on, is a level regulation planned? To what extent do wheel loads vary, must thermodynamics be taken into account? Are the spring deflections large or small? How important is the rolling behavior of the bellows? These questions already show some aspects from which the air spring system can be viewed and which have a direct influence on the model complexity and the associated parameter requirements. In the following, an attempt will be made to briefly address some of the essential points.

9.2.1 Determination of Quasi-static Stiffness

An air spring is activated under the pressure p standing gas on a surface A and thus generates the power F.

$$F = p \cdot A \tag{9.19}$$

Neither the pressure nor the surface are constant during compression and deflection. Both sizes depend on spring travel ds. In order to determine the stiffness of the spring in a first approximation, we derive Eq. (9.19) according to ds so that the following relationship is created

$$c = \frac{F}{ds} = p \cdot \frac{dA}{ds} + A \cdot \frac{dp}{ds} \tag{9.20}$$

The first part of the sum describes the change of the effective area that depends on the rolling contour and geometry of the spring. Here, a corresponding characteristic curve must be generated from the design data of the spring if the surface change contributes significantly to the change in stiffness.

The second part of the sum shows the effect of the pressure change on the stiffness of the spring. When the spring moves, the volume of the spring changes. Assuming a polytropic change of state, the following applies

$$p_1 = p_0 \cdot \left(\frac{V_0}{V_1}\right)^{\kappa} \rightarrow p \cdot V^{\kappa} = const \tag{9.21}$$

With very slow, quasi-static compression and extension movements, the heat generated by the compression of the air volume during compression can be dissipated quickly enough to the environment (isothermal, $\kappa = 1$). The change in pressure is therefore proportional to the change in volume. However, since the air spring is operated dynamically in the axle frequency range, the movement is too fast to sufficiently transfer the heat (adiabatic, $\kappa = 1.4$).[4] So the pressure rises disproportionately.

If one derives Eq. (9.21), then one obtains

$$V^\kappa \, dp + \kappa \, p \, V^{\kappa-1} \, dV = 0 \qquad (9.22)$$

$$dV = -\frac{V^\kappa}{\kappa \, p \, V^{\kappa-1}} \, dp = -\frac{V}{\kappa \, p} \, dp \qquad (9.23)$$

Describes the change in volume dV depending on spring travel ds when

$$dV = -A \, ds \qquad (9.24)$$

follows from this

$$ds = \frac{V}{\kappa \, A \, p} \, dp \qquad (9.25)$$

If this is inserted in the second part of the sum of Eq. (9.20), the result for the stiffness component due to the pressure change is as follows

$$c = \frac{\kappa \, A^2 \, p}{V} \qquad (9.26)$$

The effective area A, the volume V, the pressure p and therefore the stiffness c also depend on the deflection due to the rolling contour, so that the underlying geometry must also be taken into account here.

In addition, the harshness of the spring plays an important role, especially with small excitations. In the broadest sense, it can be regarded as a hardening of the spring, which must also be taken into account. The design of the spring and especially of the bellows (axial bellows, cross bellows, mixed forms) plays a decisive role here.

9.2.2 Determination of Dynamic Stiffness

If one considers that the gas will heat up, one must first establish the energy balance in the system, as described for example in [Scha13].

---

[4]Of course, this process is not adiabatic in the strict sense, since heat can still be dissipated to the environment, so that the isentropic exponent will be between 1 and 1.4.

$$Q + W = U \tag{9.27}$$

The sum of the thermal energy Q and mechanical work W corresponds to the inner energy U.

In the case of thermal energy, we limit ourselves to heat transfer via the outer surface A_a is released into the environment. Using the heat transfer coefficient α and the reference temperature T_0 it results from

$$Q = -\int \alpha \, A_a \, \Delta T \, dt = -\int \alpha \, A_a \, (T(t) - T_0) \, dt \tag{9.28}$$

For the mechanical work, we consider the volume change work resulting from gas compression or gas expansion. In addition, for example, the work caused by friction could also be added.

$$W = \int F \, ds = \int p \, A \, ds = \int p \, dV \tag{9.29}$$

For the internal energy, we balance the stored heat in the system with the air mass m_L and the average specific heat capacity c_m.

$$U = m_L \, c_m \, \Delta T = m_L \, c_m (T(t) - T_0) \tag{9.30}$$

Using the thermal equation of state for ideal gases and the specific gas constant for dry air R_L the air mass can be calculated in a reference condition.[5]

$$p_0 \, V_0 = m_L \, R_L \, T_0 \rightarrow m_L = \frac{p_0 \, V_0}{R_L \, T_0} \tag{9.31}$$

If one considers the temporal derivation of the balance Eq. (9.27)

$$\frac{dQ}{dt} + \frac{dW}{dt} = \frac{dU}{dt} \tag{9.32}$$

and applies for Eq. (9.29) the thermal equation of state for the pressure,

---

[5]For dry air the specific gas constant is $R_L = 0.28$ kJ/kg K. Needless to say that the temperature is of course given in Kelvin.

$$p\,V = m_L\,R_L\,T \rightarrow p = \frac{m_L\,R_L\,T}{V} \tag{9.33}$$

a first order differential equation for the temperature is obtained.

$$\dot{T}(t) + \left[\frac{\alpha\,A_a}{m_L\,c_m} - \frac{R_L\,A(s(t))\,\dot{s}(t)}{c_m\,V(s(t))}\right] T(t) - \frac{\alpha\,A_a}{m_L\,c_m}\,T_0 = 0 \tag{9.34}$$

If this differential equation is solved, the corresponding pressure can be calculated using the temperature $T(t)$ and the thermal equation of state for ideal gases and then the stiffness can be determined using Eq. (9.26) or the force of the spring can be determined multiplied by the effective area.

The structure of this equation shows that the heat transfer coefficient α, the mean specific heat capacity c_m, the course of the active surface A and the volume V over the spring deflection s are additionally needed. These are parameters which, usually, only the air spring manufacturer initially possesses. However, a good description of the suspension behavior is not possible without considering these thermodynamic effects.

9.2.3   Use of Measured Characteristic Curves

A simple approach would be to present the air spring as a steel spring with a force-stroke characteristic curve and then to switch or replace it depending on the load case (e.g. isothermal K&C test rig and adiabatic road crossing). The characteristic curves must be determined for different loads so that the different loading conditions can be taken into account. In [Puff09], concrete suggestions for measuring methods for the characterization of air springs are made.

The way to map an air spring like a steel spring with a non-linear characteristic works as long as it is only about averaged output quantities and very low-frequency excitations. Springs are often modeled as one-dimensional force elements, which only apply their forces in the connecting line of the upper and lower spring connection. The spring thus neither is subjected to forces and torques outside this line, nor can it apply them. Straight air springs are moved spatially while the wheel suspension is travelling in and out and thus inevitably applies longitudinal and transverse forces or bending moments. Exactly this force situation leads to problems when rolling the bellows during the suspension travel.

How to determine these forces is described, for example, in [Ramb08], where transverse forces of up to 20% of the vertical forces were measured. The dynamic behavior of the spring is shown, for example, in [Gaut01] or [Baue08]. In [Eich03] the modeling of air springs with additional volume is described.

9.2.4 Level Control

Vehicles equipped with air springs are often equipped with a level control system. This has two consequences. First of all, the same level must always be set in the model regardless of the load condition. Depending on the modeling type, this can be done, for example, by changing the preload of the suspension spring.

If characteristic curves are used for the air spring, a characteristic curve for the higher load and the new gas quantity must be used. With air springs, level control is achieved by changing the amount of gas until the desired level is reached.

In Eq. (9.26) the dependence of the air spring stiffness on the pressure was shown. If we now compare the state 0 (e.g. unladen) with the state 1 (e.g. laden), the result for the stiffness component due to the pressure change is as follows

$$c_0 = \frac{\kappa A^2 \, p_0}{V_0} \rightarrow c_1 = \frac{\kappa A^2 \, p_1}{V_0} \tag{9.35}$$

The quantity of gas is varied until the original volume V_0 is reached again. Due to the changed load, a new system pressure p_1 is created. The effect is that the stiffness changes proportionally to the pressure change and thus to the load change. This is an essential feature of the air spring with level control. From vibration theory, the relationship between stiffness and mass is known as natural frequency.

$$f = \frac{1}{2\pi} \sqrt{\frac{c}{m}} \tag{9.36}$$

Since the stiffness changes to the same extent as the mass, the natural frequency of the vehicle remains (almost) constant. If, therefore, only the mass is changed in the model without using an adapted characteristic curve, the vibration behavior of the vehicle body does not correspond to that of the real one.

9.3 Bound and Rebound Bump Stops

The compression and extension travel can be limited hydraulically, i.e. speed-dependent, or with springs, i.e. travel-dependent. Only the travel-dependent variants are discussed below.

9.3.1 Bound Bump Stop

Bound bump stops are used to limit the maximum compression travel and to give the spring characteristic a progressive curve (see also Sect. 9.1.1.1). Their mode of action is similar to

Fig. 9.10 Bound bump stop characteristic

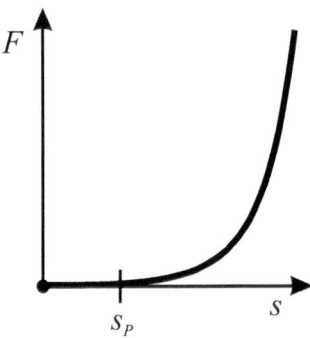

that of the elastomeric bearings described in Sect. 8.3. Normally they consist of an elastomer material and therefore have both stiffness and damping properties. The decisive difference to the bearing, which requires a different approach in terms of model technology, is that the bump stop is only used after a certain deflection.

The suspension spring has a play s_P in which it operates alone. Only when this has been completely used does the bump stop add its extra stiffness and damping (Fig. 9.10).

The stiffness of the bump stop could be defined like an additional spring using a force-stroke characteristic whose force effect only assumes an amount greater than zero when the buffer is used. However, the damping of the buffer can only be proportional to the speed if it is implemented as a simple damper model. In this way it always works, even if the bump stop is not compressed at all. This approach therefore leads to increased damping forces within the free travel of the spring.

It is more goal-oriented to represent the bump stop with the help of a contact modeling as it is provided by most simulation programs. The contact model makes it possible to specify the starting point of the bump stop and to let the laws of force take effect only from this point. Depending on the conversion, this can be done via a buffer characteristic curve and a damping constant or characteristic curve.

Contact models are often one-dimensional, i.e. care must be taken that the actual geometric conditions are converted accordingly (distance calculation).

9.3.2 Rebound Bump Stop

The rebound bump stop is intended to limit the extension spring travel. Similar to the bound bump stop, it has a play during deflection and only acts after its point of application. Linear steel springs with constant stiffness, which are installed in the damper, are often used. Since the springs have no damping effect, they can be modeled using a simple spring model with free travel.

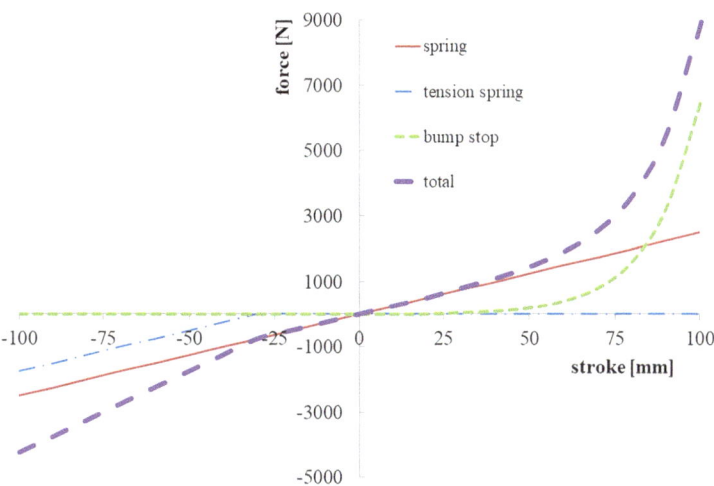

Fig. 9.11 Combination of individual component characteristic curves

9.3.3 Combination

If you model the mainspring, the bound and the rebound bump stop separately, the combination of the individual components has the effect as shown in Fig. 9.11.

If a vehicle measurement of a K&C test rig is used as a basis for the parameterization of the spring elements, it must be taken into account that the secondary spring is included in the measurement as additional stiffness. Conversely, this is a good way of determining the secondary spring if the individual component stiffness are known. The difference between the total characteristic curve shown in Fig. 9.11 and the measured wheel-related characteristic curve then corresponds to the *parasitic* secondary spring.

9.4 Spring Ratio

Depending on the modeling of the wheel suspension, the suspension spring is either mounted at the kinematically correct locations between the vehicle body and the wheel carrier or a link, as in the target vehicle, or the force is applied directly to the wheel center (see also Sect. 8.1.2). In the latter case, it should be noted that the component-related stiffness must be converted to the wheel center point.

If the spring moves from its geometrically correct position to the center of the wheel,[6] the effective lever arm and, to the same extent, the spring deflection lengthen. In order to take both effects into account, the spring ratio must be taken into account quadratically so that the same spring effect is achieved on the wheel.

---

[6]For example, because no suspension kinematics are modeled at all.

Fig. 9.12 Spring ratio

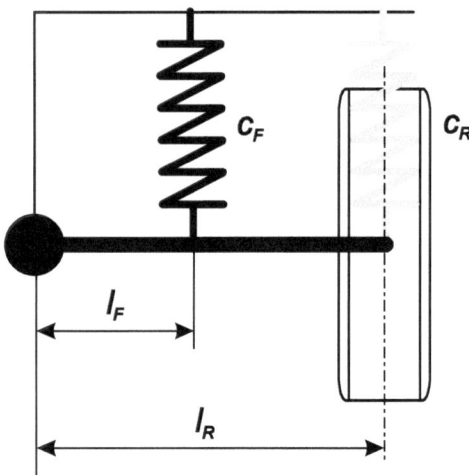

The spring ratio represents the relationship between the two lever arms l_F and l_R (Fig. 9.12). There are two ways of forming this relationship in literature and in application:
Spring ratio greater than one:

$$i_F = \frac{l_R}{l_F} \tag{9.37}$$

Spring ratio lesser than one:

$$i_F = \frac{l_F}{l_R} \tag{9.38}$$

The spring with the component stiffness c_F is to be moved from the position l_F to the wheel center position l_R. The effective lever and the spring deflection change (Fig. 9.13). The relationship between the wheel travel s_R and the spring travel s_F again represents the transmission ratio via the lever set.
Spring ratio greater than one:

$$i_F = \frac{s_R}{s_F} \tag{9.39}$$

Spring ratio lesser than one:

$$i_F = \frac{s_F}{s_R} \tag{9.40}$$

Fig. 9.13 Change of lever and
spring deflection

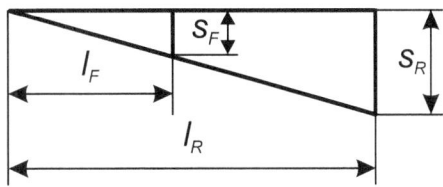

If it is assumed that the same force of the spring should act on the wheel despite the new mounting position, the use of i_F using Eqs. (9.37) and (9.39), calculate the wheel-related stiffness as follows c_R

$$l_R \cdot s_R \cdot c_R = l_F \cdot s_F \cdot c_F \tag{9.41}$$

$$c_R = \frac{l_F}{l_R} \cdot \frac{s_F}{s_R} \cdot c_F = \frac{c_F}{i_F^2} \tag{9.42}$$

Furthermore, depending on the arrangement of the spring, it should be noted that the spring ratio does not remain constant during compression and deflection because the projections of the lever lengths change. If this effect is essential, the spring characteristic curve can also be given a ratio characteristic curve i_F (s_R) or calculate them directly with the spring characteristic curve. However, two effects are then mixed so that the (converted) component characteristic curve differs from the characteristic curve used. However, four additional characteristic curve evaluations and the subsequent multiplication are saved per calculation step. In the case of very time-critical applications, this can help to reduce the computing time.

The same relationship applies to the following damper with regard to the ratio.

References

[Baue08]   BAUER, W.: *Hydropneumatische Federungssysteme*, Springer, Berlin, 2008
[Eich03]   EICHLER M. ET AL.: Dynamik von Luftfedersystemen mit Zusatzvolumen: Modellbildung, Fahrzeugsimulationen und Potenziale, in *Reifen-Fahrwerk-Fahrbahn*, VDI-Berichte 1791, Hannover, 221-241, 2003
[Fruh10]   FRUHMANN, G. ET AL: Achskonzeptstudie mit radführender Querblattfeder, in *19. Aachener Kolloquium Fahrzeug- und Motorentechnik 2010*, Aachen, 2010
[Gaut01]   GAUTERIN, F.: Noise, Vibration and Harshness of Air Spring Systems, in *Reifen-Fahrwerk-Fahrbahn*, VDI-Berichte 1632, Hannover, 273-285, 2001
[Lee03]   LEE, E.-S.: *Untersuchung des Entgleisungsverhaltens von Güterwagen mit Mehrkörpersystem (MKS)- Modell unter Berücksichtigung der COULOMBschen Reibung der geschichteten Blattfeder*, Dissertation, University of Berlin, 2003
[MeSc07]   MEISSNER, M. AND SCHORCHT, H.J.: *Metallfedern - Grundlagen, Werkstoffe, Berechnung, Gestaltung und Rechnereinsatz*, Springer Verlag, Berlin, 2007

[Puff09]   PUFF, M.: *Entwicklung einer Prüfspezifikation zur Charakterisierung von Luftfedern*, FAT-Schriftenreihe Nr. 223, VDA, Berlin, 2009

[Ramb08]   RAMBACHER, C. ET AL: Grundlagen eines sechsdimensionalen Luftfedermodells für Achssimulation, in *Berechnung und Simulation im Fahrzeugbau, VDI-Berichte 2031*, Baden-Baden, 433-445, 2008

[Scha13]   SCHARFENBAUM, I. ET AL: Modeling of air springs – required models for the virtual development process, in *4th International Munich Chassis Symposium*, 191-210, Munich, 2013

[Witt06]   WITTKOPP, T.: Mehrkörpersimulation von Schraubenfedern in der Antriebs- und Steuerungstechnik, in *Federn – Unverzichtbare Bauteile der Technik*, VDI-Berichte 1972, Fulda, 197-213, 2006

10.1 Dampers

10.1.1 Force Law and Damper Characteristic Curve

The suspension dampers used in the chassis are complex components which, nevertheless, are often only modeled as a law of force in the form of speed-proportional viscous damping (Fig. 10.1). For many simulation tasks this is also sufficient, as long as it is about *classical* single or twin-tube dampers in the *normal* operation. Since vehicle dampers are always non-linear, modeling with a damping constant, as it is used as standard in the single or dual mass oscillator approach, is ruled out for the questions dealt with here. Even the division between bound and rebound stage by a constant value of their own is not effective. Here, a strong discontinuity is generated around the zero point, which can lead to numerical problems (Fig. 10.2).

If linearization is absolutely necessary, it must be considered very carefully which working range of the damper can be mapped and whether this is really sufficient. During the bound and rebound movement of the wheel, there is a zero crossing for the damper speed at each reversal point with a subsequent sign change.

The components bound and rebound bump stop frequently fitted with the damper have already been dealt with in the previous Sect. 9.3, the top mount in Sect. 8.5.2.

The so-called VDA characteristic curve, in which the evaluation method and the evaluation points are specified, has been established as the standard (in Germany) for decades. The result of such a measurement is a force-velocity characteristic, which can be

© Springer Fachmedien Wiesbaden GmbH, part of Springer Nature 2021   183
D. Adamski, *Simulation in Chassis Technology*,
https://doi.org/10.1007/978-3-658-30678-6_10

Fig. 10.1 Electronically
adjustable damper systems

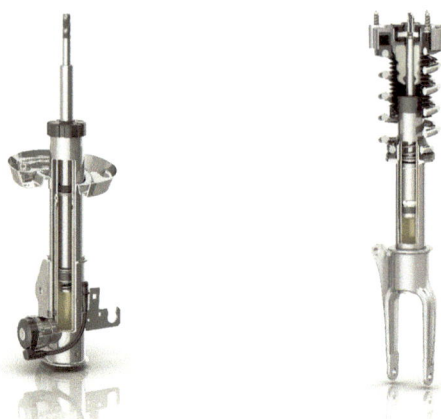

[ZF Friedrichshafen AG]

Fig. 10.2 Discontinuity with
linear transition

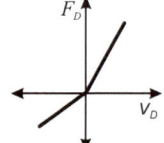

used immediately by the simple damper models.[1] The dampers are measured at constant
strokes and defined rotational or cylinder speeds or alternatively at constant speed and
adapted strokes, as shown in Fig. 10.3 as an example. The force-stroke curve shows the
work that the damper can do (Fig. 10.3, left)—this is why the term work diagram is often
used for this diagram. The larger the enclosed area of a measurement cycle, the more
energy can be converted. The speed can be easily determined from the stroke and the
rotational speed.[2] If the total stroke is $s = 100$ mm, then the damper has its maximum speed
at 50 mm, then the piston must be braked again. At a rotational speed of $n = 100$ min^{-1} is
calculated for the damper speed v_D

$$v_D = \frac{2\pi}{60\,\frac{s}{min}} \cdot n \cdot \frac{s}{2} = \frac{2\pi}{60\,\frac{s}{min}} \cdot 100\ min^{-1} \cdot 50\ mm = 0,52\frac{m}{s} \qquad (10.1)$$

In the bound and rebound movements described, damper velocities of 0 m/s are present
at each of the two reversal points. Then on the way to the other reversal point, the damper

---

[1]Here, again, the sign convention in the simulation program must be observed.

[2]The method comes from a time in which measurements were made with the aid of thrust cranks, so
that one still often finds the specification of rotational speeds in the literature. Nowadays,
measurements are mainly made on hydro pulsers.

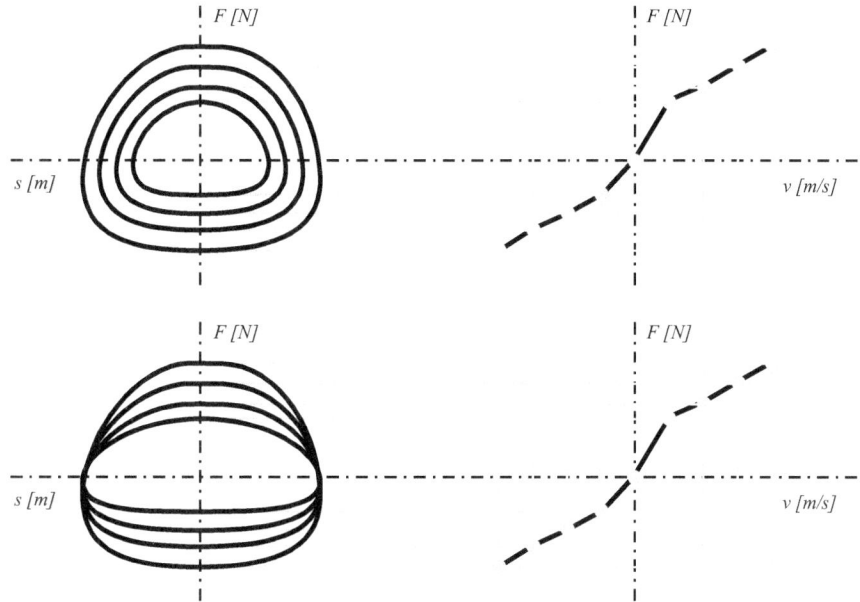

Fig. 10.3 Measurement of a damper characteristic curve with constant speed (top) and constant stroke (bottom)

accelerates to its maximum speed until it has to be braked again. In this measuring procedure, the maximum force at this stroke is evaluated and assigned to the respective damper speed.

If two dampers are compared, the quality of the outside of this maximum must be the same, since the entire characteristic is reduced to this maximum value. If the dampers differ significantly in their force over the path, as would be the case with a conventional damper compared to stroke-dependent damper, these aspects are no longer taken into account. In such cases, further data must be available and the simple model approach must be extended accordingly to include these functions.

Since this type of damper characteristic only describes the relationship between the piston speed and the resulting damper force, no effects that are dependent on the excitation frequency and/or the excitation amplitude can be mapped in these models.

The low damper speed range is interesting both for the vehicle dynamics and for the ride comfort—i.e. the behavior from the rest position ($v_D = 0$).

After the pure teaching, a force must be built up immediately for the vehicle dynamics, which reduces the body movement already in the emergence. A degressive characteristic curve comes closest to this behavior (Fig. 10.4, above). In terms of ride comfort, the problem lies in the transmission of small excitations (due to road roughness).

Ideally, no damper force should be generated in this case and thus as few disturbances as possible should be introduced into the structure and to the passengers. A progressive course

Fig. 10.4 Degressive and
progressive course of the damper
characteristic curve

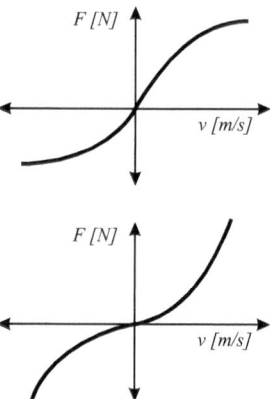

of the damper characteristic curve would be more favorable here (Fig. 10.4, bottom). Every
vehicle manufacturer solves this conflict in his own way.

The problem for the computational engineer lies in the fact that the classical measuring
method for the small damper speed range provides only a few grid points, which means that
in this important range it may be very inaccurate. In addition, friction plays a decisive role,
especially from the resting position. It must be identified in an extra measurement;
otherwise it is contained in the measured force amplitude. Therefore, an extended measuring range in the small damper speed range should be selected for the measurements and a
friction measurement should be provided.

If a measured characteristic curve is not to be used as described above, but the
characteristic curve itself is the target of the simulation, a slightly different procedure is
recommended. If an optimization is to be used to design a characteristic curve for a given
set of maneuvers, it is more advantageous not to map it using grid points, as is the case with
measurement, but section by section using straight-line equations. Figure 10.5 shows the
bound and rebound stages divided into two sections. Depending on the application, further
sections may of course be useful. For example, to be able to map the low speed range better
or to introduce a further bend at higher damper speeds.[3]

That way, the gradients of the straight lines can be used as parameters in the optimization. Suitable maneuver-specific parameters serve as target functions. As with any optimization, the mathematical solution has to be adapted to physics afterwards with the help of
engineering knowledge.

---

[3]However, the discontinuities at the transitions should be noted, as discussed at the beginning.

Fig. 10.5 Damper
characteristic curve by linear
equations

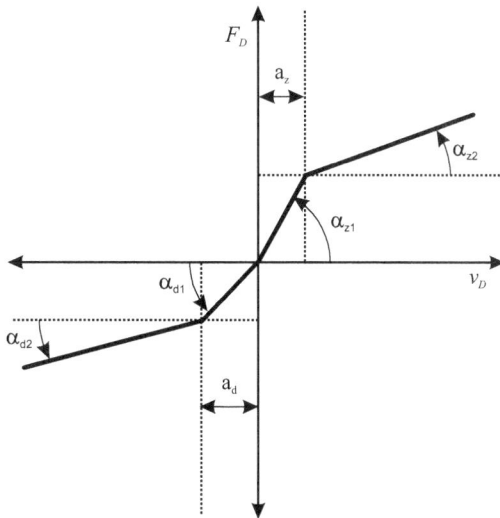

10.1.2 Kinematics and Mass

From the point of view of MBS, the damper consists of various elements. Until now, the
force element damper, which represents a force-speed function, has been treated. In
addition, the damper can act as a cylindrical joint, which is also used kinematically in
the McPherson wheel suspension concept (see Sect. 8.1.1). In addition, the mass of the
damper can be taken into account. It should be noted that most dampers are fixed to the
body on one side and to the wheel carrier or chassis on the other side. This suggests a
modeling of two bodies. The part on the wheel side belongs to the unsprung mass which,
due to its small amount, reacts sensitively to whether, for example, 4 kg is missing or not.
Of course it depends strongly on the design of the damper how much of the weight you add
to the vehicle body and how much to the wheel carrier. Where the piston rod is fixed, the
lower weight should be entered.

10.1.3 Damper Ratio

If the damper element in the model does not apply at the point intended in the target vehicle
but at the wheel center, the stored characteristic curve must be converted to this point. As
with the spring, the damper ratio must be included in this conversion in square form (see
Sect. 9.4).

10.1.4 Gas Spring Forces

Regardless of the spring elements installed on the damper, such as bump stops, the damper also delivers spring force components in addition to its damping forces. The gas pressure dampers have a gas cushion under pressure. This gas cushion provides a static basic force for the stationary damper, which is usually specified in the measurements. Since different volumes are available in the two chambers of the damper, the compressibility of the gas is used as a balancing volume. During dynamic excitation, the pressure of the gas and thus the acting spring force change due to the polytropic change of state. Nevertheless, most damper models only provide a constant offset force. The changes in the gas force are then implicitly assumed in the damper characteristic even though they are travel-dependent. However, since the current gas volume is not known in a standard measurement and can only be approximated indirectly via the pressure conditions in the damper, this path can be regarded as sufficiently accurate for the vast majority of applications.

If the damper oil is initially assumed to be incompressible, it has no elasticity and cannot store any potential energy. If it becomes compressible, spring behavior develops and a vibrating system is created. The stiffness of the liquid column is then pressure-dependent and therefore no longer constant, as was demonstrated for example in [Stre12].

10.1.5 Seals and Friction

The seals between piston and damper tube and on the piston rod (and possibly on the separating piston) have two properties which must be taken into account depending on the load case. On the one hand, friction occurs at the respective contact surface, which in the case of small excitations leads to the damper *jammed* and cannot develop any damping. On the other hand, small excitations can cause the damper to spring in the seal. This means that the elasticity of the seal allows the movement to be compensated without the piston rod moving. Of course there is no damping effect even in this case. The shape and size of the hysteresis depends strongly on the type of seal [ReSt89]. How this can be taken into account is described in Sect. 10.2. The order of magnitude of the friction can be measured in the same clamping as the damper characteristic curve. A very low speed is selected for the friction measurement, so that the damping plays a subordinate role in the measured force and the friction is predominantly measured. For the ride comfort simulation of small excitations (micro shake) these effects usually have a strong influence on the result and should therefore not be neglected or only estimated.

10.1.6 Temperature Influence

The task of the damper is to convert kinetic energy into thermal energy. Usually, these effects are not taken into account in the damper models, so that the heating of the fluid and the seals is neglected. Investigations in [LiSe01] show that the influence of temperature of the dampers between $-20\,°C$ and $+100\,°C$ causes relevant changes in the force effect. The high temperatures can occur during bad road crossings, as they are usual for the determination of loads for the fatigue strength. Whether one actually develops a thermodynamic model that can reproduce these effects or whether one makes measurements of the dampers at the temperature that can be reached and then considers them stationary depends on the maneuvers and the tasks posed. Of course, it must also be considered to what extent the real damper can be surrounded and cooled by air, depending on the driving speed of the maneuver. The preparation of a total heat balance is certainly necessary and feasible in the rarest cases.

10.1.7 Complex Damper Models

If the damper plays a prominent role for the given problem or if it is the main reason for the investigation, more attention will be paid to modeling. The damper oil is then no longer incompressible, the viscosity is temperature-dependent, the gas volume is polytropic and the valve dynamics are essential. This then requires a modeling that goes far beyond the models offered as standard. Usually, one will have to resort to programming the physical connections oneself. But despite all the enthusiasm, one must not forget that this model also has to be parameterized—and hopefully not just once. Therefore, one should always keep the simple measurability and thus the better availability of the parameters in mind. Otherwise, the model will quickly end up rotting in the back corner of the hard disk or you will only feed the supposedly accurate model with a large number of estimated parameters. A suggestion for modeling and parameterization via parameter identification can be found for example in [FuBe08].

10.2   Friction

Wherever components move, friction occurs at the contact points to the other components, in some cases on a scale that cannot be neglected.

Due to their large contact surface, friction occurs particularly in the ball joints of the linkages. The seals of the dampers also rub against the contact surfaces of the piston and the piston rod. The damper does not work until the external forces have overcome the static friction. The friction of the seals and gears also plays an important role in the steering train, especially for the steering feel. With leaf springs without plastic inserts, the friction can assume such large dimensions that additional dampers become superfluous. The friction effects are also significant with tires. However, they are so complex there that they can no longer be represented with the classical approach of COULOMB. Chapter 12 explains how this is done for tire models.

10.2.1 Coulomb's Friction

In the first semester, as a future engineer, one gets to know friction as a force opposed to movement. At this point in time, in the easily understandable, but strongly simplifying form of the Coulomb's friction. The static friction cannot be represented by the classical force elements spring or damper, because it does not represent a constant force. Rather, it blocks a degree of freedom up to a certain force level, which it releases again after exceeding the so-called breakaway force. This is not unproblematic, since most simulation programs do not provide for a varying number of degrees of freedom and this behavior is represented by an increase in stiffness. Which does not remain without influence on the computing time and the eigenvalues of the system.

First of all, a distinction must be made between static friction at standstill and dynamic friction in motion. The simplest approach is friction F_R as a force proportional to the normal force F_N and the direction of the speed v to counteract.

$$F_R = \begin{cases} -\mu \cdot F_N & \text{with } v > 0 \text{ (dynamic friction)} \\ F_H & \text{with } v = 0 \text{ (static friction)} \\ \mu \cdot F_N & \text{with } v < 0 \text{ (dynamic friction)} \end{cases}$$

Here μ is the dynamic friction value between the two components. At standstill, the holding force F_H acts as follows. It acts like a stiffening of the system and acts in the frequency domain as an increase of natural frequencies.

Numerically problematic in this modeling is the discontinuity of the force course during a change of direction—here the friction force F_R changes the sign for the same amount (Fig. 10.6). If a continuous transition is formed, this is indeed more numerically compatible, but this may lead to freedom from friction at low speeds or to a friction force which is too low, leading to incorrect results depending on the component and load case (shown dashed in Fig. 10.6). According to [Popo09], the following approach can be used for this continuous description of the friction force:

$$F_R = -\frac{2}{\pi}\,\mu\,F_N\,\arctan\,\frac{v}{v_{ch}} \tag{10.2}$$

The parameter v_{ch} represents the characteristic speed. It can be used to design the transition between static and dynamic friction (Fig. 10.7). However, the evaluation of the Arcus Tangent function is computationally time intensive.

This can be counteracted by storing the expected value range in a table and then determining the required values by linear interpolation. Another simplifying approach divides the adhesion and sliding area into two parts and approximates them by linear equations [Simp09]. The gradients m_1 and m_2 of the two straight lines are again used to calculate the friction force by comparing the current velocity at the contact point with the characteristic velocity (Fig. 10.8).

Fig. 10.6 COULOMB's friction

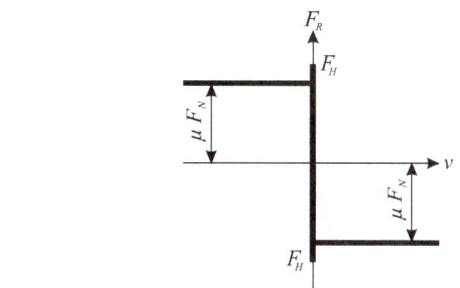

Fig. 10.7 Modeling of the COULOMB's friction according to [Popo09]

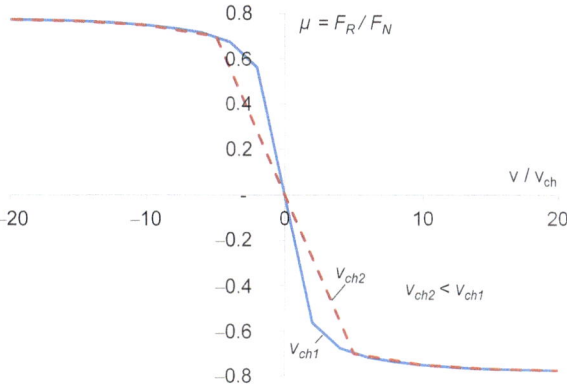

Fig. 10.8 Modeling of the COULOMB's friction according to [Simp09]

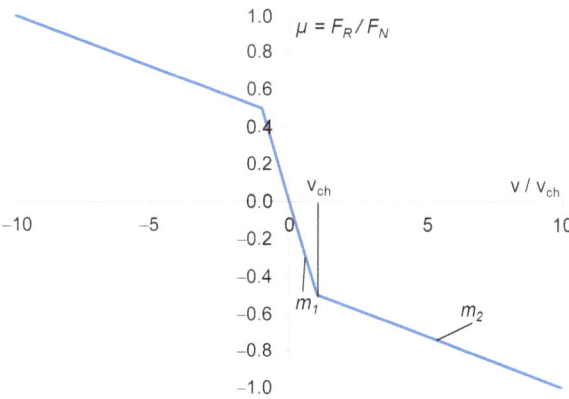

$$F_R = \begin{cases} -m_1 \cdot v_{ch} + m_2 \cdot (v_{ch} - v) & \text{with } v > v_{ch} \text{ (dynamic friction)} \\ m_1 \cdot v & \text{with } -v_{ch} \le v \le v_{ch} \text{ (static friction)} \\ m_1 \cdot v_{ch} + m_2 \cdot (v_{ch} + v) & \text{with } v < v_{ch} \text{ (dynamic friction)} \end{cases}$$

If static friction occurs only rarely in the load case under consideration, since the components are usually located in the *broken forth* state, i.e. in dynamic friction, less

Fig. 10.9 Fictitious total
friction

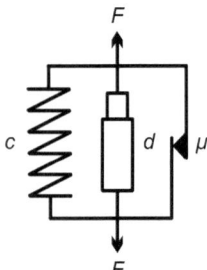

attention must be paid to this circumstance. A continuous modeling of the friction force can
be acceptable in this case. If the components repeatedly come to a standstill, the transition
from static to dynamic friction is essential and must be modeled accordingly (stick-slip
effects). However, this can lead to higher computing times, since the force law must be
switched depending on the status (stick or slip). Most simulation programs offer more
complex friction models for this purpose, which, however, require a larger number of
parameters in addition to the longer computing time. Special measurements are necessary
to parameterize these models. My recommendation here is to first use the simple models
and understand their limitations before attempting the complex models.

10.2.2 Fictitious Total Friction

The consideration of the friction effects on each component in the chassis leads to
extremely long computing times and is rarely useful. However, it is also rarely possible
to completely neglect them, as otherwise a too optimistic picture would be presented. An
alternative can be a fictitious total friction, which is installed at one point (e.g. parallel to the
suspension spring or damper) and summarizes all friction effects in the vertical direction
(Fig. 10.9). The order of magnitude of this total friction can be measured on a K&C test
stand, as the force path between the wheel center and the vehicle body is available there.
Maneuvers on very even roads (e.g. micro shake) benefit from this type of modeling.
However, the effects of friction in the longitudinal or lateral direction cannot be reproduced
in this way.

References

[FuBe08]   FUNKE, T. AND BESTLE, D.: Modellierung, Parameteridentifikation und Simulation
            passiver Fahrzeugstoßdämpfer, in *Simulation und Test in der Funktions- und
            Softwareentwicklung für die Automobilelektronik II*, Expert Verlag, Renningen,
            130-140, 2008

[LiSe01]   Lion, A. and Sedlan, K.: Anwendungen spezieller Komponentenmodelle zur Simula-
            tion von Fahrwerkbelastungen auf Schlechtwegstrecken und Komfortanalyse, in *Reifen-
            Fahrwerk-Fahrbahn*, VDI-Berichte 1632, 225-253, 2001
[Popo09]   Popov, V. L.: *Kontaktmechanik und Reibung*, Springer, Berlin, 2009
[ReSt89]   Reimpell, J. and Stoll, H.: *Fahrwerktechnik: Stoß- und Schwingungsdämpfer*, Vogel
            Verlag, Wuerzburg, 1989
[Simp09]   Simpack: *Force Element Catalogue*, Manual, 2009
[Stre12]   Stretz, A.: *Komfortrelevante Wechselwirkung von Fahrzeugschwingungs-dämpfern
            und den elastischen Dämpferlagern*, Dissertation, University of Darmstadt, 2012

Steering

In this chapter, only the steering assembly is described (Fig. 11.1). The influence of the driver, the origin and implementation of the steering request, i.e. the steering control, is described in Sect. 16.2.

The driver's wish for steering can be transferred to the steering system via two different physical variables. The first thing to think about is the steering wheel angle. In most cases it will also be the input of the steering model (Fig. 11.2). The steering model converts this steering wheel angle δ_H in the case of a two-track vehicle, kinematically into a wheel steer angle of the outer wheel δ_a and the inner wheel δ_i. Whether this happens directly or via a steering gear, which in turn has, for example, a rack shift as an output, is of secondary importance at this point. The wheel movement results in a wheel side force F_R which in turn acts back on the steering wheel and there in a required steering torque M_H results. Other parameters such as friction, servo support and tire reset torque, etc. are initially bundled in this force F_R.

Since the steering wheel angle is often specified as a function of time (e.g. via a value table), its derivatives, in particular the steering wheel angle speed, are implicitly specified. This should be kept in mind, as the speed at which the driver steers should be appropriate to the maneuver and the modeled driver. For standard maneuvers, such as steer step input according to [ISO7401], these speeds are specified (200–500°/s).

But also the possibility to alternatively specify the steering torque makes sense (Fig. 11.3). In combination with a steering control for the driver, the maximum torque to be applied by the driver can be limited very well and thus a realistic driver behavior can be modeled.

First, the usual question has to be asked again, what the steering model is used for. If it is only used for following the course of the vehicle, simple model approaches, as described in the following section, may suffice. If it is important that the required steering torque

© Springer Fachmedien Wiesbaden GmbH, part of Springer Nature 2021
D. Adamski, *Simulation in Chassis Technology*,
https://doi.org/10.1007/978-3-658-30678-6_11

Fig. 11.1 EPS steering of a
Mercedes SL (R231)

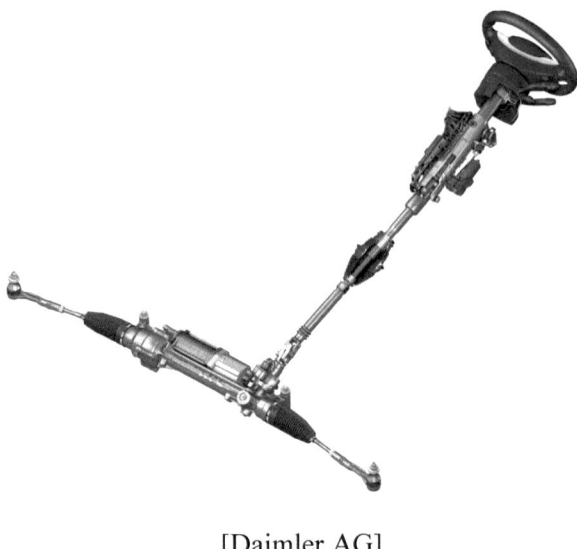

[Daimler AG]

Fig. 11.2 Steering wheel angle
as input of the steering model

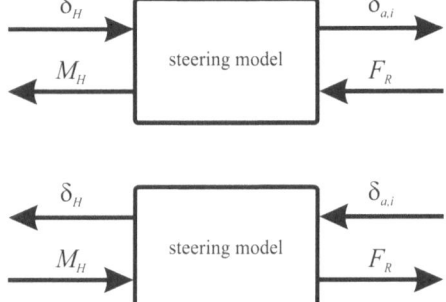

Fig. 11.3 Steering wheel
torque as input of the steering
model

including the feedback of the road or the stress of individual steering components can be
mapped, more effort must be put into the modeling. Such models are explained in the
following sections.

11.1   Simple Steering Models

The simplest representation of the steering is certainly the restriction to the corresponding
degree of freedom on the wheel, which is directly influenced by a wheel steer angle
specification. With this approach, the steering is not modeled as itself. The *driver's wish*
of the steering control is directly related to the steering movement of the individual wheel

or axle. It should of course be noted that the inner wheel needs a different steer angle than the outer wheel (Fig. 11.7). If there are no manufacturer-specific specifications, modeling according to the Ackermann conditions is certainly the first step in determining the required steer angles on the two wheels (Fig. 11.5). For this, you first need the position of the center of gravity in x-direction.

With a given wheelbase l and the weight distribution in the vehicle (front axle load m_v, total vehicle weight m_{ges}), the distance between the center of gravity and the rear axle l_h can be determined with Eq. (11.1) (Fig. 11.4).

$$l_h = l \cdot \frac{m_v}{m_{ges}} \tag{11.1}$$

Now you can geometrically determine the Ackerman angle δ_A from Fig. 11.5 with Eq. (11.2). Assuming sufficiently large curve radii, the Ackermann angle becomes correspondingly small and the equation can be linearized.

$$\tan \delta_A = \frac{l}{\sqrt{R^2 - l_h^2}} \rightarrow \delta_A = \frac{l}{R} \tag{11.2}$$

As with any linearization, the range of values in which the operation is performed must be kept in mind. If the investigated curve radii are smaller, then larger angles are created and the error due to linearization increases.

Furthermore, the static definition of the Ackerman angle is based on the assumption that no slip angles occur on the wheels. By definition, this is not possible with cornering where a lateral force must be transmitted. The resulting slip angles at the front and rear axles cause the instantaneous center of rotation to shift from the center of the circle, resulting in a deviation here as well (Fig. 11.6).

In order to now be able to adjust the wheel steer angle of a two-track vehicle (Fig. 11.7) to the wheel steer angle of the inner (δ_i) and the outer wheel (δ_a), you have to set the track to b_v of the front axle (Eqs. 11.3 and 11.4).

$$\delta_i = \frac{l}{R - \frac{b_v}{2}} \tag{11.3}$$

$$\delta_a = \frac{l}{R + \frac{b_v}{2}} \tag{11.4}$$

Fig. 11.4 C.O.G. position

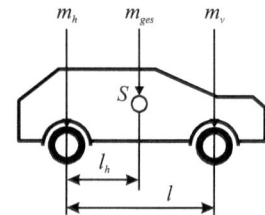

Fig. 11.5 Static Ackerman
angle

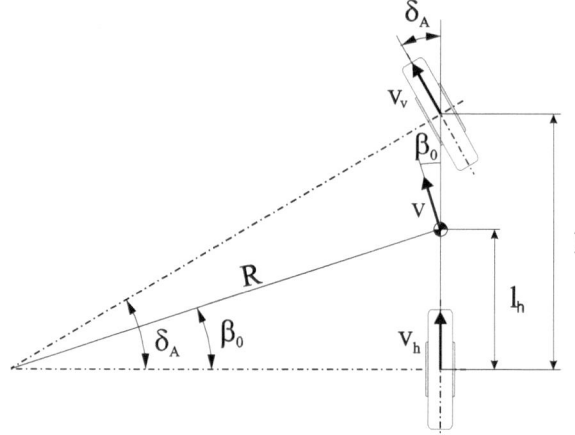

Fig. 11.6 Dynamic Ackerman
angle

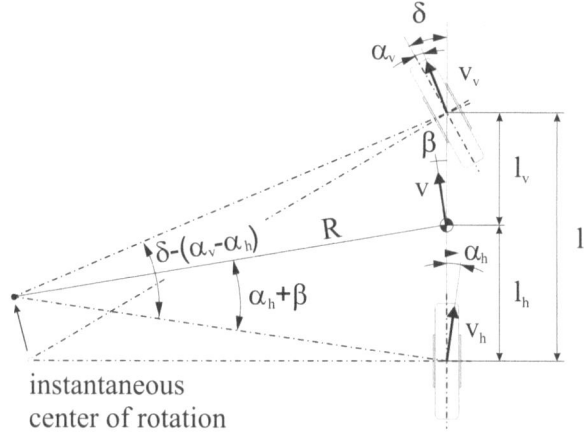

instantaneous
center of rotation

A maximum angle should limit the turning of the wheels. Here, too, if no vehicle-specific specifications are available, an initial estimate can be made on the basis of the track circle diameter or the turning circle. If you still do not know anything about the body overhangs and mirror widths of the vehicle, you can put the track circle diameter and the turning circle diameter equal in a first approximation.

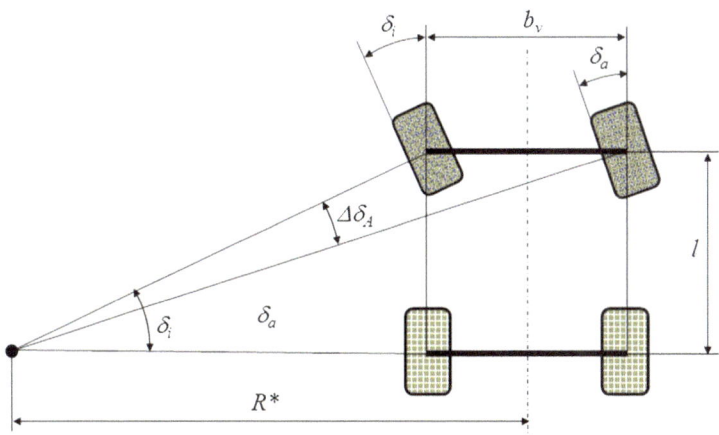

Fig. 11.7 Required wheel steer angles for a two-track vehicle (With sufficiently large curve radii, the center distance of the rear axle to the instantaneous pole $R*$ can be equated with the actual curve radius R)

In the equations used so far, no scrub radius is taken into account. In [Mats07] it is proposed to use the so-called steering pin track, which represents the distance of the intersection of the spreading axis with the horizontal plane through the wheel center. Accordingly, the track b_v must be replaced by this value.

Statements about the required steering torque cannot be made with this type of modeling.

11.2   Steering Train

11.2.1 Steering Gear

The function of the steering gear is to represent the steering ratio i_L between the steering wheel angle δ_H and the mean wheel steer angle δ_m.

$$\delta_m = \frac{\delta_a + \delta_i}{2} \tag{11.5}$$

$$i_L = \frac{\delta_H}{\delta_m} \tag{11.6}$$

As an example, at a steering ratio of 1:15, a steering wheel angle of 15° shall produce an average wheel steer angle of 1°. First of all, the type of steering gear involved (rack and pinion steering, recirculating ball gear, etc.) is neglected. Depending on the simulation environment, this translation is performed either by a constraint or a special gear element.

It is important that the ratio is signed. This is due to the fact that it depends on whether the steering gear engages the track rod in front of or behind the axle center (and how this circumstance is taken into account in the modeling). With rack-and-pinion steering, this can easily be achieved. If the driver steers to the left, the vehicle should of course also drive to the left. If the steering gear is in front of the axle center in the direction of travel when viewed from the driver, pull the left front wheel inwards and push the right front wheel outwards (Fig. 11.8). If the gearbox is located behind the center of the axle, the conditions are reversed. For steering models that can be used with several different axle concepts and variants, as required by modularity, this must be taken into account. The check can be done by a simple circular drive, so you can immediately see what a positive and what a negative steering wheel angle causes. According to [DIN70000], a positive steering wheel angle steers to the left.

Racks are increasingly being used that have a variable transmission ratio over the path, which in vehicles with power steering becomes more indirect in the center area and outwards on both sides more direct, in order to reduce the steering effort when parking (Fig. 11.9).[1] This can be stored as a characteristic curve over the rack travel in the constraint or the gear element.

For some vehicles, the customer can determine the transition between indirect and direct transmission by choosing a standard steering system or a so-called direct or progressive steering system. For this reason, it is advisable to make the necessary characteristic easily replaceable by storing it in a file or database. If the ratio is to be coupled directly to the steering wheel angle, it should be noted that there are elasticities between the steering wheel and the gear. They can lead to a relative angle between the steering wheel and the pinion.

Friction occurs between the pinion and the rack, which can be taken into account in the gearbox model or added globally to a fictitious total friction in the steering train, depending on the model.

11.2.2 Steering Column

As can be seen in Fig. 11.10, today's steering columns are no longer continuous but divided into three parts. Two universal joints allow a bent course of the steering column. The greater the bend angle of the universal joints, the greater the uneven transmission of the steer angle—also known as steering ripple (Fig. 11.11). Sometimes, the steering ripple is consciously used, so that it should also be found in the model. This can be done by

---

[1]For vehicles without assistance this is exactly the opposite. The central area is as direct as possible for the necessary straight-line driving and the outer area becomes more indirect. The steering angle increases, but the necessary steering torque is reduced.

Fig. 11.8 Position of the gear

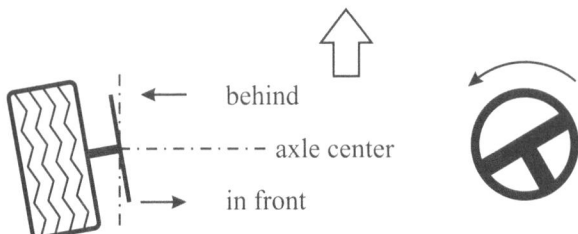

Fig. 11.9 Variable steering ratio over the rack and pinion stroke

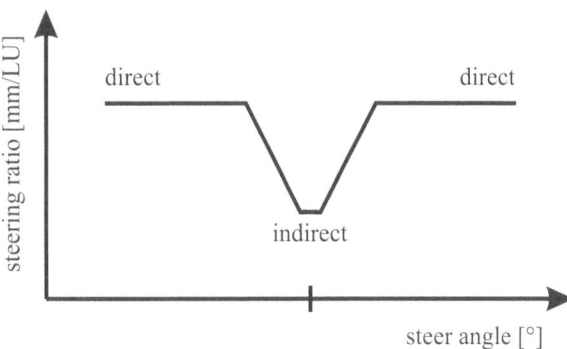

modeling the steering column with two universal joints or by superimposing the sinusoidal waviness on the steering ratio.

The difference angle $\Delta\alpha$ from the input angle α_1 and the output angle α_2 can be determined depending on the bend angle β of the universal joint.

$$\Delta\alpha = \alpha_2 - \alpha_1 = \mathrm{atan}\left(\frac{\tan \alpha_1}{\cos \beta}\right) \tag{11.7}$$

The position of the second universal joint in relation to the first one then makes it possible to compensate this difference angle again or to use the influence on the steering ratio for steering from the center.

If the vehicle is offered in a left-hand drive and a right-hand drive version, the angles of curvature are sometimes very large, so that constant velocity joints are used instead of universal joints, which have no angular error [PfHa13].

The elasticity of the steering column is essential for the vibration behavior. It can be modeled using weakly damped torsional stiffness. To do this, the steering column has to be cut open and reconnected with a revolution joint and a torsion spring. If a Hardy's disk (hard rubber disk) is installed, its stiffness and damping can also be taken into account in the torsion spring.

Fig. 11.10 Steering column
with rack

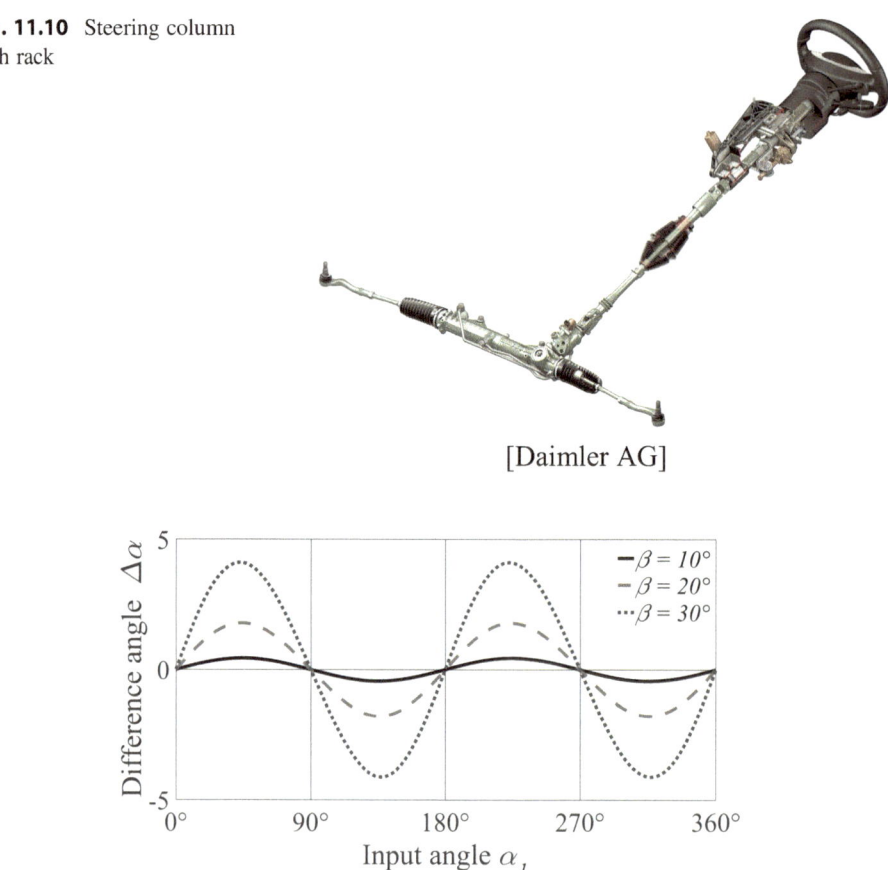

[Daimler AG]

Fig. 11.11 Angular error of a universal joint with bend angle β

11.2.3 Steering Wheel

For the consideration of steering wheel vibrations, it is necessary to map the entire steering train up to the steering wheel in its stiffness and damping behavior as well as its mass properties. This also includes the mass properties of the steering wheel, in particular the moment of inertia for the torsional vibrations. Of course, it has to be considered if a mass damper is installed in the steering wheel, which has to be modeled if necessary.

If the diameter of the steering wheel is also taken into account, the required manual force can simply be determined from the steering torque or a force instead of a torque can be specified as the driver's steering request.

11.3   Power Steering

Once the mechanical steering train has been modeled, the question is whether assist power steering has to be displayed in the model if it is present in the vehicle. If it is only a question of lane guidance, then this is certainly not necessary. However, many of today's assistance systems, which were made possible by the introduction of electric power steering (EPS), cannot be fully simulated without considering this.

The power assistance must be clearly separated from additional assistance functions in terms of its effectiveness, particularly in the case of electric power steering. The supporting force or torque may only be supplied if the hand torque exceeds a certain value. In the real system this is determined by the torsion of the torsion bar. In most cases, the torsion bar must be modeled independently of the power steering concept. Its rotation usually serves as an input to the power steering model. It should be noted that the maximum torsion of the torsion bar is limited by a stop. Superimposed functions such as vehicle speed-dependent force control can make additional inputs necessary.

11.3.1 Hydraulic Power Steering (HPS)

The torsion of the torsion bar due to the supportable steering torque causes a displacement of the valve openings and ends the short-circuit delivery of the servo oil. The oil can now flow into the corresponding chamber of the hydraulic cylinder and generates a supporting force across the piston surface depending on the applied pressure. In the simplest case, you can now calculate the moment at the torsion bar M_{DS} with the torsion angle of the torsion bar γ_{DS} from its torsional stiffness c_{DS}.[2]

$$M_{DS} = \min\left(c_{DS} \cdot \gamma_{DS}, M_{DS,max}\right) \qquad (11.8)$$

The pressure p_{zyl} can be converted via a characteristic curve over the surface A_{zyl} in the right or left cylinder chamber to the supporting force F_{HPS} (Fig. 11.12). Depending on the vehicle, the required torque can vary with the vehicle speed v.

$$F_{HPS} = p_{Zyl}(M_{DS}(\gamma_{DS}), v) \cdot A_{Zyl} \qquad (11.9)$$

When implementing power steering, the causality must be taken into account (see also Sect. 2.1.2). The supporting force can only become effective if there is torsion of the torsion bar due to a torque supported between the steering wheel and the tire. In a real vehicle, this

---

[2]To protect the torsion bar, its maximum twist is limited by a stop. This is considered in the following Eq. (11.8) with a maximum torque $M_{DS,max}$.

Fig. 11.12 Vehicle speed
dependent characteristic map of
a hydraulic power steering
system

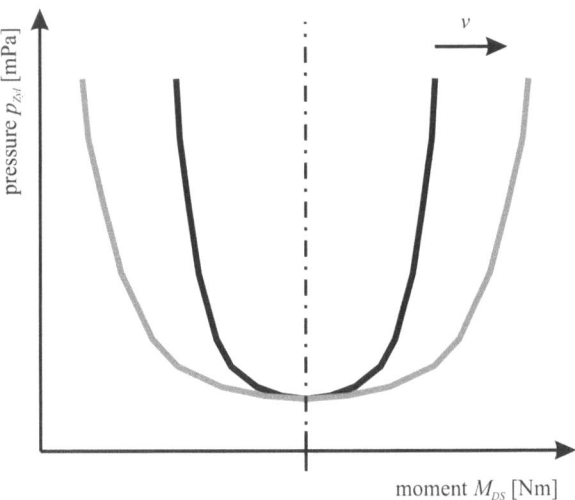

relationship prevents unintentional steering. In the model, the calculation of the supporting
force must also be coupled to the torsion bar, otherwise the force may act at the wrong time.

Time behavior, such as hardening of the steering, cannot be mapped in this way. For
this, it must be known how long the pressure and thus the force build-up takes. This can be
important for highly dynamic maneuvers. For this reason, the flow rate and performance of
the pump must also be taken into account. The pump is suspended in the belt drive of the
combustion engine. It is therefore dependent on the current engine speed. If no
corresponding engine model is used that calculates the correct motor speed depending on
the load situation, it can be regarded as constant for specific load cases. For highly dynamic
driving maneuvers such as step steer or the lane change test, a typical engine speed can be
inferred from the vehicle speed and the transmission ratio. The biggest challenge is parking
at idle speed. Here you can check whether the respective delivery volume is sufficient when
the engine is idling.

To prevent the servo oil from escaping, appropriate seals have been installed, which
provide additional friction due to the system pressure.

11.3.2 Electrohydraulic Power Steering (EHPS)

The hydraulic connections between the torsion bar controlled valve and the working
cylinder are the same as for the hydraulic power steering. The decisive difference between
the HPS and the EHPS lies in the control of the pump. While in the HPS it is coupled
directly to the speed of the combustion engine, in the EHPS it is driven by its own electric
motor. That way, the flow rate of the pump can be controlled depending on the driving
situation. The corresponding control strategy must be known. It can then either be stored in
a control algorithm of the model and thus the pump can be correctly controlled in all

driving maneuvers or, as described in the last section, a suitable power for this driving situation is specified as constant. If the power consumption of the electric drive is to be investigated, a corresponding model of the electric motor must be available.

11.3.3 Electric Power Steering (EPS)

In the case of electrical power steering, a distinction must first be made between where and how the support torque of the electric motor is applied. Depending on the wheel loads and thus the maximum rack forces required, the motor/gear unit is attached directly to the steering column (EPSc) (Fig. 11.13), to the pinion (EPSp) or to the rack (EPSrc, EPSapa) (Fig. 11.14 or Fig. 11.15). Since packaging has no relevance in the model, the elasticity of

Fig. 11.13 EPSc steering of the VW up!

[Volkswagen AG]

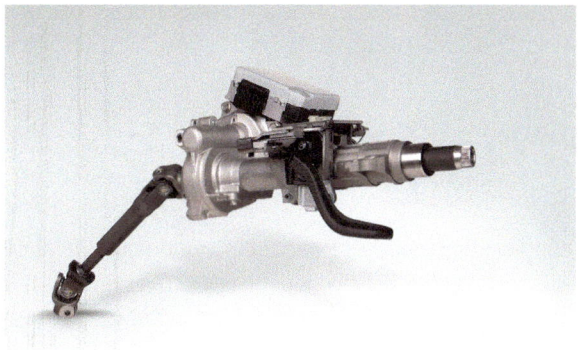

[NSK Europe Ltd]

[Audi AG]

Fig. 11.14 EPSrc steering of the Audi A6 (C7)

the steering train is of particular importance. If these are to be taken into account, the steering column must be modeled to be torsional soft. In addition to the stiffness, a Hardy's disc also provides the material damping of the elastomer material.

Fig. 11.15 EPSapa steering of
the Mercedes M-Class (W166)

[Daimler AG]

References

[DIN70000]   DIN 70000: *Straßenfahrzeuge, Fahrzeugdynamik und Fahrzeugverhalten, Begriffe*,
             1994
[ISO7401]   DIN ISO 7401: *Testverfahren für querdynamisches Übertragungsverhalten*, 1989
[Mats07]   MATSCHINSKY, W.: *Radführungen für Straßenfahrzeuge*, Springer Verlag, Berlin,
             2007
[PfHa13]   PFEFFER, P. AND HARRER, M. (ED.): *Lenkungshandbuch*, Springer Vieweg, Wiesbaden,
             2013

„Without tires we'd all be traveling in streetcars or on railroads, and none of us could live more than a walking distance from a rail line. Cars, trucks, bicycles, motorcycles – none would be possible. That would be a different world!" [Hane03]

In order to understand the vehicle behavior when accelerating and braking, when cornering and when transmitting road unevenness or surface roughness, it is necessary to know how forces and torques arise in the tire and how they are transmitted (Fig. 12.1). The contact surface between the tire and the road, the so-called contact patch, is essential. Here, depending on the material properties of the tire and those of the road, it is decided which forces can and cannot be transmitted. One must always keep in mind that the motion of the vehicle is based on frictional connection and that this can be overstrained if the force is too great, which leads to slip and thus to skidding.

The tire is therefore one of the most important components of the chassis, because the contact patch is the only contact with the road and most measures in the chassis area aim to keep this patch as large as possible. GILLESPIE describes this very clearly in [Gill92]:

„It has often been said that the primary forces by which a high-speed motor vehicle is controlled are developed in four patches – each the size of a man's hand – where the tires contact the road. This is indeed the case."

Without the tire, no transmission of longitudinal and lateral forces and no support of the vehicle weight would be possible. It has a strong influence on the driving behavior and accordingly the care must be great when selecting a suitable model approach and the necessary parameters. In addition to its mass properties, which are often combined in a rigid rim within the MBS model, it is above all the transmission of force and torque that is the essential property of a tire model. Depending on the complexity of the model, this *force element* is part of the simulation environment or added as part of a co-simulation. If you see

© Springer Fachmedien Wiesbaden GmbH, part of Springer Nature 2021
D. Adamski, *Simulation in Chassis Technology*,
https://doi.org/10.1007/978-3-658-30678-6_12

[Continental AG]

Fig. 12.1 SUV tires (left) and a section of the Contidrom (right)

the tire model as a transmission element, you can see the position and orientation of the wheel center and its derivatives as well as the road information (elevation profile, coefficient of friction, etc.) as inputs. They then produce the forces in the longitudinal, lateral and radial directions as well as the corresponding torques (Fig. 12.2).

For long-wave excitations there are often approaches which consist of only a spring or a spring/damper combination and can therefore only represent the vertical suspension behavior of the tire, supplemented by a slip model. In many approaches of this kind, the damper is dispensed with, since, so the common argument, the tire damping is of small order compared to the suspension damping. This is often also acceptable in combination with the suspension damper, at the latest if you want to represent the tire characteristics alone (tire test bench), this approach is no longer sufficient. However, this saves the differentiation of the road signal in these simple models.

In the literature, the classification into

- linearized models,
- models based on approximated maps,
- simple deformation models and
- structural models

is often found. In general, the models can also be distinguished by the fact that, on the one hand, they try to represent the physics of the tire or, on the other hand, they try to describe

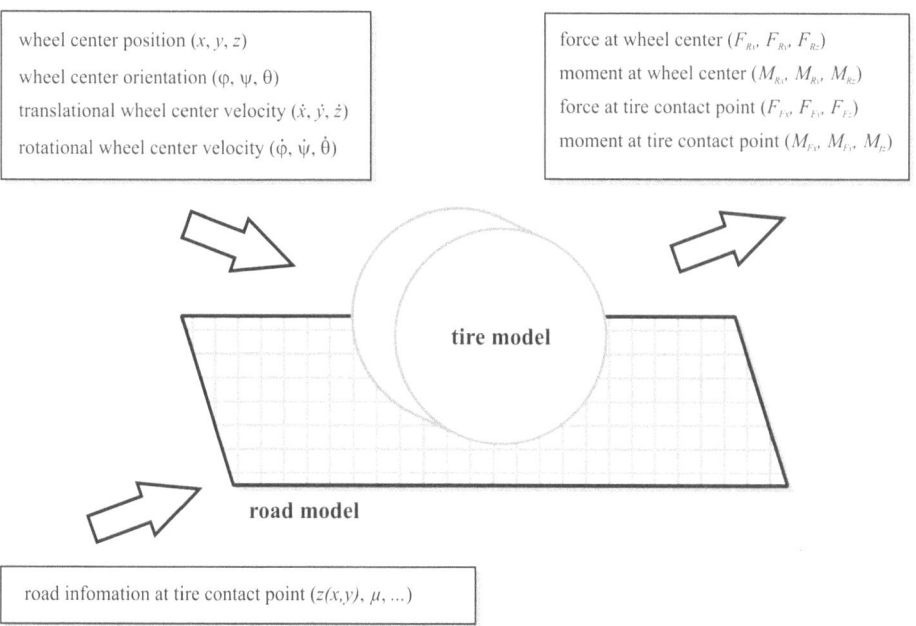

wheel center position (x, y, z)

wheel center orientation (φ, ψ, θ)

translational wheel center velocity $(\dot{x}, \dot{y}, \dot{z})$

rotational wheel center velocity $(\dot{\varphi}, \dot{\psi}, \dot{\theta})$

force at wheel center $(F_{R_x}, F_{R_y}, F_{R_z})$

moment at wheel center $(M_{R_x}, M_{R_y}, M_{R_z})$

force at tire contact point $(F_{T_x}, F_{T_y}, F_{T_z})$

moment at tire contact point $(M_{T_x}, M_{T_y}, M_{T_z})$

tire model

road model

road infomation at tire contact point $(z(x,y), \mu, ...)$

Fig. 12.2 Transmission element tire model

the behavior of the tire with mathematical approximation formulae. The physical models include, for example, the FTire described in Sect. 12.3.1 or the RMOD-K presented in Sect. 12.3.2. The best known mathematical, non-physical model is the Magic Formula (Sect. 12.2.1).

Finally, the tire model calculates the contact forces in the contact patch. Whether the contact patch in the tire model represents a point, a line or a surface depends on the depth of the modeling and whether a short-wave road surface is to be scanned or not. The contact forces are generated by normal and shear stresses due to deformation caused by longitudinal forces (braking, driving), lateral forces (cornering, steering) or vertical forces (uneven road surface, wheel load fluctuations).

Structural models, which are calculated with the aid of implicit or explicit FEA solvers, are generally only used for tire design by the tire manufacturer. For reasons of computing time and parameter requirements, MBS models are more likely to be used for vehicle design, but some of them take approaches from the FE models. Section 12.3.3 presents a tire model in which FE formalisms are used in an MBS model.

If you want to dive deeper into the kinematics between the tire and the road, you cannot get past [Pace06]. The important role the tire model plays in the reconstruction of vehicle accidents is described in detail in [BuMo09].

Table 12.1 Typical requirements for tire models in the MBS environment

	Frequency	Amplitude	Longitudinal slip	Lateral slip
Vehicle dynamics	<5 Hz	Low	Medium to high	Medium to high
Vehicle dynamics control systems	<5 Hz	Low	Medium to very high	Medium to very high
Ride comfort	<30 Hz	Low to high	Low	Low
Load data determination	<30 Hz	Very high	Low to medium	Low to medium

12.1   General Requirements for Tire Models

Of course, the requirements placed on tire models are again strongly dependent on the issues in which they are used. In the field of vehicle dynamics, ride comfort or load data determination, the excitations of the road differ in amplitude and frequency content. The longitudinal and lateral slip values also vary depending on the application. Table 12.1 shows a selection of typical requirements that may vary within a discipline due to different load cases.

Although the necessary vehicle dynamics knowledge is assumed at most points, some basic terms must be repeated for tire modeling so that the following explanations can be understood.

12.1.1 Modeling the Contact Patch

The contact area relevant for the transmission of the force of the tire is the tire contact patch. In models that are used exclusively for vehicle dynamics, this area is usually not determined from the actual deformation of the tire, but approximated. This model type is therefore only suitable for long-wave ground unevenness. In [OeEF99], a ratio of 2.5% between patch length and wavelength of the roadway is given as a rule of thumb and a ratio of 10:1 is given for the relationship between road curvature and tire radius. The contact between the tire and the road surface here is related to an idealized wheel contact point, so that unevenness within the patch cannot be detected at all. The decisive advantage of this model class lies in the low computing time, which makes these models real-time capable.

If shorter-wave excitations, i.e. road roughness, potholes or cleats on test benches are to be taken into account, the patch must be described more accurately. For this, it is necessary to model the structure of the tire in such a way that the pressure distribution can be calculated from the deformation of the tire. This requires a detailed modeling with many degrees of freedom—the advantage lies in the more exact scanning of the road—the disadvantage in the higher computing time. Such tire models are often the determining factor of the computing time and bring the vehicle models at a great distance from the real-time capability.

If individual obstacles are to be crossed, the quality of the calculation results depends decisively on the scanning of the obstacles. Only if the contour of the obstacle is sufficiently accurately captured by the contact surface during the crossing can its force effect on the tire be correctly reproduced (see Sect. 12.5.3).

12.1.2 Friction Contact and Slip Definition

The friction force acting between the tire and the road consists of two components. The adhesion between the boundary layer of the rubber and that of the road has the largest portion with dry road. Here, binding forces act on the molecular level. The second, smaller proportion is caused by the hysteresis of the rubber. If a rubber element is deformed when it enters the contact patch, part of the energy input is dissipated by rubber due to the high material damping. This means that less energy is available when springing back, an effect known as hysteresis. Especially on wet roads, the coefficient of friction between tire and road drops drastically. The adhesive forces now act between the tire and the water where there is no longer any direct road contact. Extensive studies on the behavior of tires on wet roads can be found, for example, in [Schr02]. The special features of large-dimension tires for off-road commercial vehicles are described in [Dudz05]. Parameters for a simplified model for different soil qualities are also given there.

Classically, the coefficient of friction between the road surface and the tire is called the μ-value. Since this coefficient of friction is anything but a constant, its change behavior must be mapped, for example, via slip in the tire model (Fig. 12.3). The specified coefficient of friction then corresponds to the maximum coefficient of friction or the coefficient of static friction depending on the modeling (in this case, the sliding coefficient of friction at 100% slip can be specified as the second parameter). The previously described effects of adhesion and hysteresis are therefore not mapped directly, but are implicitly contained in the specified coefficient of friction. When simulating on a wet road, this value is reduced accordingly compared to the dry road. If only the coefficient of static friction is specified for the model, the course of the μ-slip curve is scaled the same over the entire slip range. The stronger slump at 100% slip (complete gliding) is not taken into account in this way. For this, at least the coefficient of sliding friction is required as additional information.

This procedure is sufficient for most maneuvers, at least as long as it is assumed that the coefficient of friction for the respective tire does not change during the maneuver. However, if you want to use a roadway with different coefficients of friction, as is the case, for example, with the μ-split braking (Fig. 12.4), the assignment of the coefficient of friction to a tire is no longer sufficient. Here, the coefficient of friction must be assigned to the road.

Both solutions, the coefficient of friction in the tire data set and the coefficient of friction in the road data set, are not ideal. Ultimately, the coefficient of friction always depends on the tire/road pairing, which means that the same tire on another road may give a different

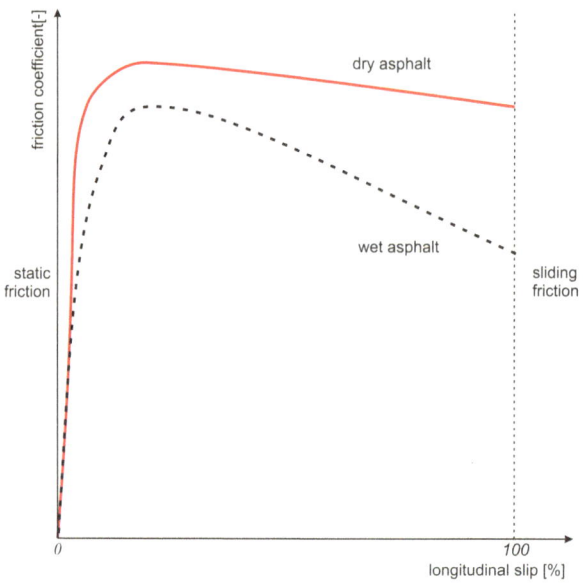

Fig. 12.3 Course of the coefficient of friction over the slip

[Continental AG]

	20 m	10 m	10 m	10 m	20 m
2 m	$\mu = 0.9$	$\mu = 0.2$	$\mu = 0.9$	$\mu = 0.2$	$\mu = 0.9$
2 m	$\mu = 0.2$	$\mu = 0.9$	$\mu = 0.2$	$\mu = 0.9$	$\mu = 0.2$

Fig. 12.4 Coefficients of friction with μ-split

Fig. 12.5 Pure rolling

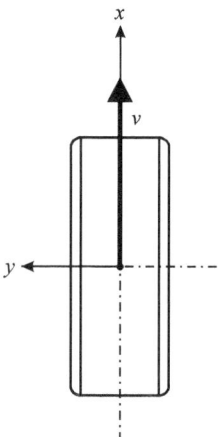

coefficient of friction and two different tires on the same road may give the same coefficient of friction. In general, the choice of tire model is a decision that has already been made on the use of the coefficient of friction, but some models also allow the user to choose where the source of the coefficient of friction lies.[1]

For pure rolling without lateral forces, the tire is by definition slip-free, i.e. there are no relative speeds between the vehicle speed and the longitudinal and lateral speed of the tire contact patch. This is a theoretical limit, but must be included in the slip definitions. We are talking about definitions here, since there are several ways to describe slip.

First of all, the transition from rolling (non-slip) to sliding (slip) must be considered. In the first case, the movement is parallel to the wheel plane (Fig. 12.5).

In this case, however, neither longitudinal nor lateral forces can be transmitted. If the tire starts to slip on the road, forces occur in the direction of travel. Let us first consider a lateral slip of the tire to the wheel plane (Fig. 12.6). The extreme case, that the angle between the direction of travel and the wheel plane is $90°$, can be illustrated by a vehicle slipping out of the curve in the direction of the curb. There is no longer any longitudinal movement. The angle between the wheel plane and the direction of travel (i.e. the direction of the resulting velocity vector) is called the slip angle α. The so-called lateral slip represents the relationship between the lateral velocity $v_y = v \cdot \sin \alpha$ and the resulting speed v and can be described using the following equation:

---

[1]There are models that calculate the longitudinal forces with a different coefficient of friction than the lateral forces. This then requires two coefficients of friction. Physically, this is critical, but allows a simpler *tuning* of the tire data set. The physical relationships should always be kept in mind. Otherwise you will quickly have a well-fitting data set for the one measurement at hand, which may compensate for one error with another and lead to a physically absurd combination.

Fig. 12.6 Side slip

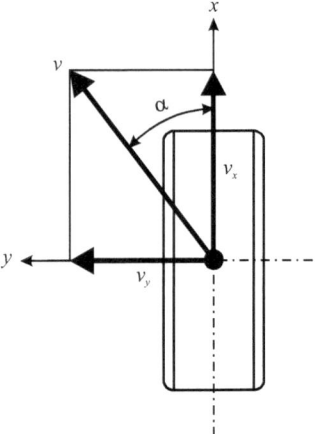

$$\lambda_y = \frac{v_y}{v} = \frac{v \cdot \sin \alpha}{v} = \sin \alpha \qquad (12.1)$$

A minimum knowledge of the kinematic determination of angular velocities is required for longitudinal slip. On the tire we can determine the speed at two points (Fig. 12.7). At the center of the wheel M the approximate vehicle speed prevails v if one assumes an almost rigid connection.

If one takes the point for the rotatory wheel movement A as the instantaneous pole, the velocity is located at the wheel center as the product of the wheel rotation (angular velocity ω) and the dynamic tire radius r. In the slip-free case of rolling, the following therefore applies to the wheel center point $v = \omega \cdot r$ and for the point A is $v = 0$, because for the point A there is no gliding.

Once sliding occurs at the point A (braking or accelerating) v and $\omega \cdot r$ are no longer identical. The definition of the longitudinal slip is not always the same in the literature, but a distinction is often made between drive and brake slip:

Brake slip:

$$\lambda_{x,B} = \frac{v - \omega \cdot r}{v} \qquad (12.2)$$

Drive slip:

$$\lambda_{x,A} = \frac{\omega \cdot r - v}{\omega \cdot r} \qquad (12.3)$$

That way, the borderline cases rolling ($\lambda_{x,A} = \lambda_{x,B} = 0\%$) and for braking locking ($\lambda_{x,B} = 100\%$) and for accelerating the spinning wheel when the vehicle is at a standstill ($\lambda_{x,A} =$

Fig. 12.7 Longitudinal slip

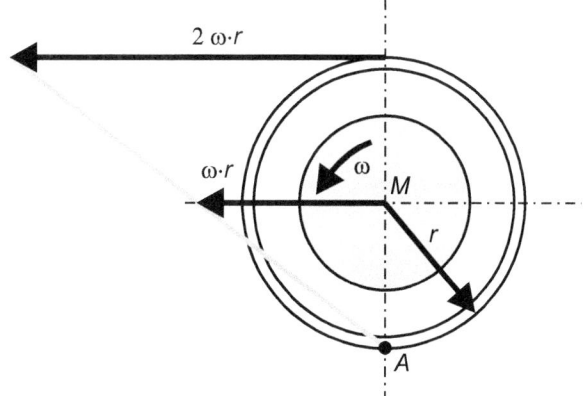

100%) are determined without getting a sign change. This type of slip definition only applies to $v > 0$ respectively $\omega \cdot r > 0$ and can therefore not be used when the vehicle is at a standstill. The problem of standstill is addressed again in the following.

12.1.3 Limits of the Slip Definition

12.1.3.1 Standstill

In the case of maneuvers that are to start from a standstill or lead to a standstill, it must first be clarified whether the tire model used can also cover this case. In the low speed range, the forces and torques due to tire deformation in the contact patch play an increasingly important role compared to the forces due to longitudinal slip and slip angle. The slip definitions described above do not cover this effect.

The limit value of Eq. (12.2) goes to one when braking with a vehicle speed of zero, i.e. according to this definition, sliding occurs. This means a force can be transmitted that would lead to a vehicle movement, at least to an *unnatural* vibration behavior. The relaxation processes in the tire contact patch at low speeds and the static friction effects are not taken into account in these simple considerations. For some models, this effect is approached by *additional support*. The extent to which these are acceptable for the own load case must be tested in advance.

12.1.3.2 Different Speed Directions

There may be cases where the direction of travel of the vehicle and the rotational speed resulting from the direction of rotation of the wheel do not point in the same direction. For this you need a slope if you do not want to overtax the drive train. A horror scenario for

Fig. 12.8 Sliding on slopes
with spinning wheels

every car driver is the situation of standing on a slippery road on a slope at a red traffic light.[2] I offer two possibilities of horror:

1) In a panic, you press the accelerator pedal too ambitiously when the traffic light turns green (Fig. 12.8). The wheel spins and due to the remaining sliding friction ($\mu_G < \mu_H$), the slope downforce cannot be overcome—the vehicle slips downhill (vehicle direction backwards, wheel direction forwards).[3]
2) One stands happily at the red traffic light and suddenly notices that the vehicle begins to slip away, because the slope downforce is stronger than the static friction force to be applied (vehicle direction backwards, wheel stationary, Fig. 12.9). After that, case 1 often occurs.[4]

If these scenarios belong to the reader's portfolio, the behavior of the tire model should be made plausible before productive use and of course validated later, but we already covered that topic.

12.1.4 Standard Tyre Interface

Based on tire workshops in Germany, the first version of a standard for data exchange between simulation programs and tire models was developed in the early 1990s. Later, an international working group turned it into the *Tyre Data Exchange Format* (TYDEX) and the *Standard Tyre Interface* (STI) [UnZa97]. Most commercial simulation programs support this standard interface. Usually, however, the programs also offer their own proprietary interfaces, which have been optimized for the data transfer of the respective tire model and thus require significantly less computing time. If one wants to adapt one's

---

[2]It is described for optimistic reasons that things are looking uphill. The downhill version works the same way, but with different signs (backwards on a slope upwards).

[3]I know that there is a traction control system. But simulation of passive vehicles has not gone out of fashion yet.

[4]Case 3 (vehicle direction backwards, wheel direction backwards) is not mentioned here, because it should fit the slip definition. However, checking the tire model you may have written yourself for negative signs could sometimes be appropriate.

Fig. 12.9 Sliding on slopes
with standing wheels

own tire model to such an environment, one must consider at this point that an unfavorably
programmed data exchange will lead to significantly higher computing times.

12.2   Tire Models for Vehicle Dynamics

As discussed before, most vehicle dynamics maneuvers take place on an ideally even
road—at least in simulation. During test drives, the road surface is of course no longer ideal
and often neither even nor smooth in everyday traffic. Nevertheless, a class of tire models
has been developed for vehicle dynamics simulation, which places particular emphasis on
the longitudinal and lateral slip display with the necessary vertical deflection. A scanning of
the road unevenness does not take place there, so that only correspondingly long wave
(or ideally even) roads can be used. However, this has an immense computing time
advantage, so that this limitation is by no means to be seen as a disadvantage—unless
you want to simulate vehicle dynamics on real road profiles. If, for example, a trip through
the Nordschleife of the Nürburgring (Germany) is to be simulated, the elevation profile of
the trajectory can certainly be mapped. However, one has to do without the unevenness of
the road as well as the driving over of the curbs.

The deformation of the sidewall is not taken into account in this model class. The
roadway and rim are regarded as rigid bodies coupled to a rigid tire spring. The contact
takes place at an idealized tire contact point where the integral contact force is applied. An
extension of this model class are models with a rigid ring. Here, the self-aligning torque
between the ring (belt) and the rim can also be taken into account.

The lateral forces that are decisive for vehicle dynamics (F_y) and circumferential force
(F_x) are essentially determined by the wheel load (F_z), the slip angle (α), the camber angle
(γ), the wheel angular velocity (ω), the longitudinal wheel speed (v_x) and the coefficient of
friction μ of the tire-road contact.

The use of a linear relationship between cornering stiffness and slip angle is limited to
the very small linear range and does not adequately reflect the real behavior of lateral force
transmission.

12.2.1 Magic Formula

Probably the world's best-known tire model, the Magic Formula was created in the mid-1980s as a result of a collaboration between University of Delft in the Netherlands under the leadership of Prof. PACEJKA and Volvo [Pace06]. The basic idea behind this model is to map the behavior of the tire (e.g. self-aligning torque over the slip angle) by a mathematical function, i.e. phenomenological (Fig. 12.10). The function that originally served as the basis was

$$y = D \cdot sin\,[C \cdot arctan\,(B \cdot x - E \cdot [B \cdot x - \ arctan\,(B \cdot x)])] \tag{12.4}$$

where $Y(X) = y(x) + S_V$ is the required output variable (e.g. self-aligning torque) corrected with the offset S_V and X the input variable (e.g. slip angle) (with $x = X + S_H$). The coefficients B, C, D and E ultimately represent the shape of the curve, in which B represents the stiffness, C the shape, D the maximum value and E the curvature factor. The factors S_H and S_V allow the curve to be shifted from zero. The parameters of the model provide the opportunity to represent a variety of waveforms and thus many tire measurements of characteristic quantities can be mapped by adjusting these factors. Figure 12.11 shows this as an example for the factor B, which represents the stiffness, in a range from 0.1 to 1.0. The other parameters were kept constant ($C = 2, D = 1, E = 0.95$).

Due to the easy modification possibility of the formulas and conversions used in the model, there are now various *dialects* to the original Magic Formula model (the formulas can be found in [Pace06]) or to the current TNO version (the description follows in the next section). In [ReRL07] no less than ten different dialects are mentioned. This means that before exchanging model data, it must be made clear that the model level is comparable and if not, where the differences lie.

Since no physical effects are mapped in the model, the tire must be real and sufficient measurements must have been carried out. How the variation of a physical parameter affects the driving behavior therefore cannot be represented.[5]

12.2.2 MF-Tyre and MF-SWIFT

TNO in Delft is a research institute where the Magic Formula tire model described above was developed in close cooperation with University of Delft and Prof. PACEJKA. The work of PACEJKA is now continued by TNO and the tire model in its original form is primarily designed for longitudinal and lateral dynamics, marketed as MF-Tyre and distributed as an

---

[5]Unless you have such extensive experience that you know the effect of the variation on the fitting parameters of the tire model.

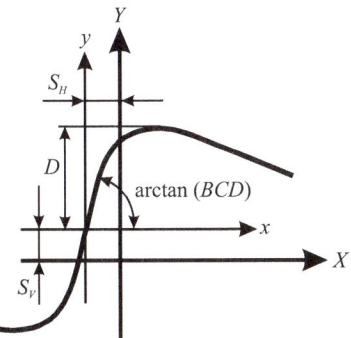

Fig. 12.10 Basic formula of the Magic Formula

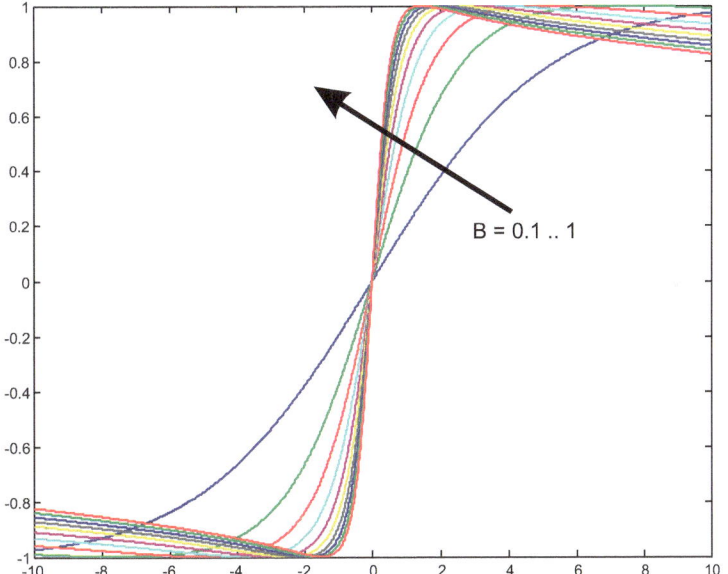

Fig. 12.11 Magic Formula: variation of the parameter B (stiffness)

integral part or extension for many commercial simulation tools. The TNO indicates the working range of the model with excitations up to 8 Hz.

An extension of the basic model for uneven road surfaces under consideration of belt dynamics was developed at the beginning of the 1990s as a cooperation between TNO and University of Delft under the name MF-SWIFT (*Short Wavelength Intermediate Frequency Tyre*) [ScBJ07]. Initially, the focus was on requirements from the development of vehicle dynamics control systems; at the end of the 1990s, the focus was increasingly on the calculation of ride comfort and load data prediction [Zege98]. Thus, the scanning of three-dimensional roads is now implemented. Implementations are available for all major commercial MBS simulation packages, which are either licensed by TNO or by the program provider.

The model consists of a rigid belt combining the mass and inertia properties. The natural vibrations in the belt can therefore not be reproduced. In the three main directions, the belt with spring-damper elements is attached to the rim so that the relative movement between rim and tread can be represented. The road surface is scanned by several elliptical discs to generate a substitute contact area from this information, which is used to calculate the forces and torques in the tire contact area. The slip calculation is carried out using the Magic Formula.

12.2.3 HSRI-Model

This model dates back to the 1970s and was built at the former *Highway Safety Research Institute* (HSRI), today's *Transportation Research Institute* (UMTRI) of the *University of Michigan at Ann Arbor* (USA). It belongs to the class of simple deformation models.

The modeled tire consists of a non-deformable carcass to which the contact patch is elastically attached. In the model no camber angle is considered and a constant surface pressure is assumed for the entire contact patch. A deflection of the contact patch relative to the carcass only occurs when a shear stress is applied (Fig. 12.12).

The centerline of the carcass is deflected relative to the wheel center plane and the contact patch centerline is deflected relative to the carcass centerline by a constant value. These simplifications are due to the age of the model, which was originally designed for analog computers and therefore the complexity and the resulting computing time had to be considerably reduced. From today's point of view, this is no longer as mandatory, therefore, before applying this model, it must be checked how relevant these restrictions are for the issues to be investigated.

If the vehicle is moving at the speed v and the wheel with the angular velocity ω, then any point P in the contact patch will move relative to its inlet point 0 in the time Δt with the

Fig. 12.12 Displacement between carcass and tire-road contact patch

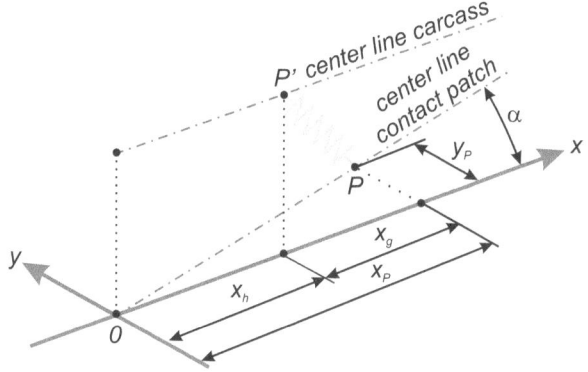

displacement $x_p = v \cdot \Delta t$. With pure rolling the corresponding carcass point P' would be the same x-y-position (the contact patch is subject only to static friction). If we assume that longitudinal and lateral slip are to be taken into account, the sliding of the point P initiates a shear stress, which causes a displacement between the P and P'. The carcass is still subject to the rolling condition and will be moved $x_h = \omega \cdot r_{dyn} \cdot \Delta t$ in Δt. The displacements x_g and y_p between the two points can be determined with knowledge of longitudinal slip λ and the slip angle α using Eqs. (12.5) and (12.6).

$$x_g = x_P - x_h = v \cdot \Delta t - \omega \cdot r_{dyn} \cdot \Delta t = \frac{v - \omega \cdot r_{dyn}}{v} v \cdot \Delta t = \lambda \cdot x_P \qquad (12.5)$$

$$y_p = x_P \cdot \tan \alpha \qquad (12.6)$$

That way, it is possible to quantify the size of the adhesion and sliding areas in the contact patch. For both areas, the calculation of the occurring stresses must be carried out separately and the resulting forces can then be determined by means of integration over the entire contact patch. In [DuFS69], this procedure is described for the original model. Extensions of the model by the self-aligning torque and the consideration of wheel load fluctuations can be found in [Wieg74] and in [Uffe80].

12.3   Tire Models for Ride Comfort and Load Prediction

In this class of applications, the transmission of forces and torques in all three spatial directions is of course also important. In addition, complex three-dimensional road profiles are to be scanned and the resulting vibrations transferred from the model. The offer of such models is relatively small, which shows how difficult it is to create such a tire model and to parameterize it correctly.

This model class contains mappings of parts of the physical properties of the tires so that, within certain limits, parameter variations can be made to determine the influence on driving behavior. For example, in most models it is possible to change the inflation pressure within a certain range around the nominal pressure.

The minimum requirement is a discretized contact surface. This is capable of scanning road unevenness in the relevant wavelength range and thus calculating the forces arising from the deformation in the tire and the short-wave excitation. Depending on the application, several contacts must be able to be processed simultaneously. If the high-frequency behavior of the tire is also of interest, the structure of the tire must be mapped. These models differentiate between the tire structure and the tread, which are linked by massless spring/damper elements in the axial, radial and tangential directions. Ground and/or sensor points are then stored in their nodes. That way, the original rigid belt is discretized. The sum of the resulting forces and torques then acts on the rigid rim, which usually also represents the interface between the MBS vehicle model and the tire model. A detailed description of this complex can be found, for example, in [OeEF99], [Oert11] or [Gips01].

In the tread, the individual cleat is represented by a contact, on which friction forces act, which result from the local sliding speed and the local coefficient of friction. The resulting individual forces or deformations are then used to calculate a total force or total deformation, which is passed on to the rim.

Few models can map the penetration of the tire onto the rim. In addition to the technical representation of the model, the parameterization is very complex and difficult. That is the boundary between MBS and FEA calculation, since it is a highly dynamic maneuver, but a plastic deformation of the rim should be taken into account.

12.3.1 FTire

The tire model FTire (*Flexible Ring Tire Model*) by Prof. GIPSER (Hochschule Esslingen) was developed in 1998 from his tire models DNS-Tire (*Dynamic Non-Linear Spatial Tire model*) and BRIT (*Brush and Ring Tire model*) ([Gips99], [Ammo95]). The DNS-Tire model consists of a coarse meshed FE net. It runs in the time domain and therefore has a correspondingly high computing time requirement. It is in a revised version as FETire part of the FTire model family. A model with a rigid belt (RTire) completes the model range. Today, it is distributed by the company Cosin.

The individual belt elements are connected to the rigid rim via friction elements, which control the hysteresis and MAXWELL-elements, which represent the dynamic stiffness of the rubber. A non-linear spring element takes over the static force transmission. In addition to the STI interface to the simulation program described in Sect. 12.1.3, FTire uses its own proprietary implementation called CTI (*Cosin Tire Interface*).

In addition to a temperature model, a model to consider tread wear, misuse can be calculated. Here, you can differentiate between ideally rigid and flexible rims. There is a separate program for the identification of tire parameters from tire measurements. FTire is offered for the common commercial multi-purpose simulation platforms and vehicle dynamics simulation packages [GiHo14].

A comparison of the tire models MF-SWIFT and FTire for the application of load data prediction can be found in [Wase07].

12.3.2 RMOD-K

This model was originally developed as a commissioned work by Volkswagen AG at gedas in cooperation with University of Berlin and University of Applied Sciences Anhalt with the aim of obtaining a model family suitable for various applications [OeEF99]. Version 6.0 distinguishes between three variants of different performance capabilities. The RFN20 model with discretized contact surface was offered as a pure vehicle dynamics model, while the RFN30 (single-track model) and RFN31 (multi-track model) were offered for ride comfort and durability. These models were later transferred to the former CDTire family.

The new development RMOD-K 7.0 is now available as Open Source at www.rmod-k.com. The basic structure of the model is documented in detail in [Oert11]. Here you can understand very well how a tire model has to be constructed to meet the requirements of ride comfort and durability calculations.

12.3.3 CDTire

In cooperation with the Fraunhofer Institute (LBF) in Darmstadt (Germany) and the company LMS International[6] in Leuven (Belgium), a set of physical tire models was first developed based on the code of RMOD-K 6, which also contains models of three complexity levels [LMS09]. The simplest model (model 20) has a rigid belt. The other two models have a flexible belt and can either only accommodate uneven surfaces in the x- (model 30) or combined in x- and y-direction (model 40) [GaBä07].

In the meantime, however, a completely new development has emerged both for a real-time vehicle dynamics model and for a complex 3D model at the Fraunhofer Institute for Industrial Mathematics ITWM in Kaiserslautern (Germany). The real-time variant corresponds to a model with a flexible, mass-produced belt and contact modeling via brushes, which is based on the previous model CDTire30, but was revised for real-time capability and ported from Fortran to C [Burg13]. The complete revision of the tire model's internal integration process and the communication between tire model and vehicle model in terms of co-simulation played an important role. The model exists in a version for hard real-time conditions as CDTire/Realtime and in an offline version as CDTire/HPS (High Performance Solver).

The CDTire/3D version has been given a completely new design that is more oriented towards the geometry and structure of the tire and reflects the sidewall and individual layers of the tread, belt plies and carcass [GaBB14]. If necessary (and if this information is available), individual steel belt layers with their thread angles can be taken into account. Large deformations, as they occur with misuse load cases, can be mapped with appropriate parameterization. The contact model is also based on a brush model. A separate tool is provided for identifying the tire parameters from tire measurements.

Both tire models are offered for the common commercial multi-purpose simulation platforms.

---

[6]Now Siemens PLM Software.

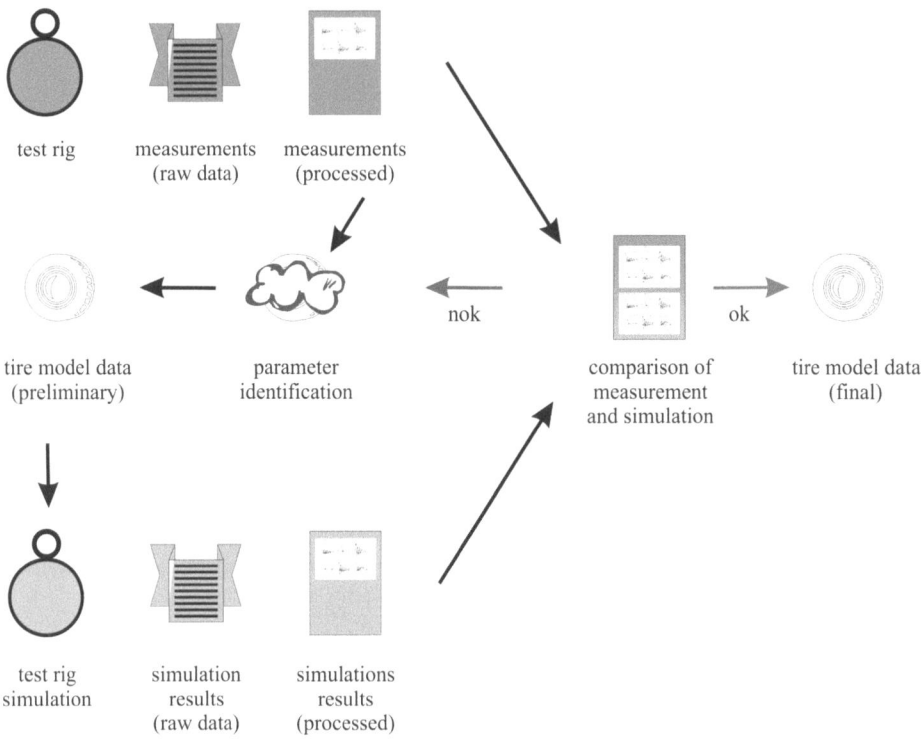

Fig. 12.13 Basic process of tire parameterization

12.4   Parameterization of the Tire Models

Even though the producers of tire models naturally see it differently, I see less the choice of the individual model as most important,[7] but the process-safe procurement of the tire parameters and the knowledge of the limits of the model.

12.4.1 Parameterization Process

Figure 12.13 shows a typical tire parameterization process. First, the target tire is measured on a tire test rig of the tire manufacturer or a service provider. Which parameters are measured and which tests are carried out varies and is mainly determined by the tire model and the associated requirements with regard to the later statement.

---

[7]The right model class should of course already be chosen.

This selection is essential for the later quality of the data. However, since the existing know-how is very widely distributed in this area, a comprehensive coordination between tire manufacturer, tire model manufacturer and the executing computational engineer would be necessary. Experience shows that this is usually not feasible. The TMPT project discussed later in this section attempts to make a suggestion here. Once the measurement data is available, a tire model data set is created using parameter identification. As most tire models are characterized by a large number of parameters, only a few *essential* parameters are often used for identification purposes to minimize the error between simulation and measurement. That is where it gets tricky. The person who carries out the identification is rarely the person who will later carry out the simulations. In which question and with which emphasis the tire data set is to be used later, is perhaps only rudimentarily known (or not at all). The parameter identification is often carried out by the tire manufacturer, the test rig operator or an external service provider, so that a clear agreement on the later use of the data or the model would be absolutely necessary. If you are lucky as a user of this tire model, you will receive the underlying measurement data with the tire data set. Now you can check the quality and usability of the data for your own task. However, this step is often omitted, either because the required measurements were not supplied or simply because there is no time. There is a large quality deficit here, which can only be reduced by a suitable process and binding agreements between the parties involved.

The keyword *process-safe* is often used in the automotive industry to demonstrate consistent quality standards. That is easier said than done in this context. Usually, the situation is such that the combination of different tire manufacturers with various products on different test rigs with a variety of measurement and identification methods can lead to a difficult to compare number of model parameter sets. That this can even come to a certain variety, with a set of tires of the same manufacturer, which were measured on different test rigs and whose characteristic parameters were generated, was the motivation of the 1996–1999 running TIME project (*Tire Measurements*) to create a uniform measuring procedure for the lateral forces [Klaa99]. But even there, the variance of the measurement results was still too big.

As a result, the University of Vienna awarded the *Tire Model Performance Test* (TMPT). With the cooperation of several vehicle, tire and tire model manufacturers and providers of MBS simulation programs, standardized measurement methods are to be defined. A detailed description of the methodology and procedure can be found in [LuPl07]. A tire (205/55 R16 90H) was measured and the measurement results were made available to tire model manufacturers for parameterization of their models. These parameterized models were passed on to the providers of the MBS programs, who passed on the simulations carried out with them to the Vienna University of Technology so that they could be collected and published there. For capacity reasons, not all combinations of tire models and simulation packages were performed for all defined maneuvers. Here, too, a relatively large dispersion of the results can be seen, which in part also showed that the integration of a tire model into different simulation environments can have an influence on the result.

external drum   internal drum   flat track

Fig. 12.14 Test rig concepts for the measurement of the rolling tire

12.4.2 Measurement of Tire Parameters

First, a distinction must be made between whether the target tire already exists, whether a comparable tire exists or whether there are only target values according to which the tire model is to be parameterized. In the latter case, the tire model used should be as simple as possible to allow a rough description based on the few targets. Under certain circumstances, some of the parameters can be obtained from an existing FE model.

If an existing tire is to be measured, several methods are available to obtain different classes of parameters. For the vibration behavior in the higher frequency range, modal analyses from the calculation or the test can be used as a basis. Test rig or road measurements are used for the dynamics of the rolling tire. Three basic concepts can be distinguished for the test rig (Fig. 12.14), which differ above all in the contact patch between the tire and the test rig. It is concave for the external drum test rig and convex for the internal drum test rig, both of which do not correspond to the contact patch of the tire on the road. The class of flat track test rigs tries to remedy this situation. However, controlling the flexible belt at high wheel loads is a challenge here. The most common are certainly the external drum test rigs, as they are the easiest to implement. Usually, the drive and most of the drum are located in the cellar and only the used part of the drum is visible. On the drum, impact bars (*cleat*) can be attached in order to be able to measure the impulse response of the tire. Due to the centrifugal force, measurements are not possible on wet roads or with snow.

Internal drum test rigs are ideal for this purpose, which can also be equipped with real road surfaces. As can be seen in Fig. 12.15, the dimensions are significantly larger in contrast to the external drum test rig. The shown drum of the test rig vehicle/roadway of the Federal Highway Research Institute in Bergisch-Gladbach, Germany, fills an entire hall. It weighs 40 tons and has a diameter of 5.5 m.

The procedure for generating model parameters from measurements on a flat track is described, for example, in [ScFö09].

Alternatively, you can also measure on real roads. Here, the general conditions such as the coefficient of friction and the temperature are not as constant as in a rig test, but the tire-road contact is clearly more comparable to the later use of the tire on the vehicle. Special vehicles or trailers are required for the measurement so that typical settings such as different wheel loads, camber or slip angles can be made during the measurement.

Fig. 12.15 Internal drum test rig of the Federal Highway Research Institute (bast)

[bast]

12.4.3 Models of Varying Complexity

Some tire models offer designs of varying complexity, so that both vehicle dynamics and ride comfort models can be offered within a model family. The parameterization of these models is naturally different, i.e. at least two data sets must exist for a concrete tire. If both data sets now refer to the same tire (manufacturer, model, tire size), the user is suggested that it is actually a replaceable model. Care must be taken here to ensure that this claim can also be met. Comfort oriented tire models should also be able to cover the linear part of a steady-state circular test and provide comparable results to vehicle dynamics models if the same tire is described.

12.5 Modeling the Road

The results of the full vehicle simulation must always be validated on measurements of real vehicles. The simplest and most convincing way to do that is to use the same road information as for the vehicle measurements as the basis for the simulations. It is therefore advisable to have at least a selection of the routes used by the driving test available in digital form. In some cases, the creation of the necessary road data is offered by service providers, in others, that is carried out by vehicle manufacturers or suppliers themselves using simplified procedures.

12.5.1 Measurement Methods for Road Profiles

There are many methods of measuring how real road profiles can be digitized and thus made available for simulation. A frequently used method is the recalculation of the z-profile of a road from measured acceleration signals. This procedure can be implemented quickly with relatively little effort. You need an acceleration sensor on the wheel carrier and the corresponding measuring equipment. By integrating the acceleration signal twice, a path signal is obtained that still has to be filtered and cleaned. On the one hand the vertical movement of the vehicle is contained in the signal, on the other hand the signal sometimes drifts, which can lead to an undesirable rise or fall of the road. Even if one measures on the right and left side of the vehicle, in the end, only the respective driving lane is measured. The information where this lies on the width of the road is lost in the process. In [Kudr01], a measurement of the road over only two measuring points on the vehicle is described, with the help of which the profile can be generated after deduction of the vehicle movement. With this method, however, an inclination of the road cannot be recorded and only the respective lanes of the measuring vehicle are measured. For carriageways without any significant inclination and clearer lane, that is an inexpensive method to measure test tracks, but also public roads easily and quickly. If it is only about the qualitative description of a street profile, for example as a general representation of a bad country road, that procedure is completely sufficient. However, it is rather unsuitable for the validation of calculation results with vehicle measurements, since the lane cannot always be reproducibly maintained on public roads, especially in curves.

If only one lane is measured, it is then run over by both tires of an axle, so that no throwing[8] can be brought in. The identical profile thus extends over the entire width of the road, so that only one piece of information $z(x)$ must be queried (Fig. 12.16, left). If two lanes are measured (Fig. 12.16, center), different excitations are possible for the left and right wheel, which will also cause a throwing. The interpolation between the two lanes depends on the street model. If the middle area is crossed due to a course deviation, this has however nothing to do with the measured road. In the right representation in Fig. 12.16, four measured tracks can be seen, but depending on the measurement method, of course there can be more. Here, the middle area of the lane is shown in more detail. Especially in curves, the lane is strongly dependent on the driving speed and the driver model (see also Chap. 16), so that deviations from the measurements due to the road excitation can occur.

A more complex method for measuring the road surface is described in [GiAR05]. There, a laser scanner mounted on a measuring vehicle is used to record the entire width of the vehicle and, after the vehicle movement has been subtracted, it is made available in digital form. The advantage of this method is that the course of the road between the actual lanes is also recorded and can thus also be used for vehicles with different track. Since the measuring vehicle has an inertial measuring technique, inclinations and absolute height differences of the road surface can also be taken into account.

Methods which photographically capture the road, such as stereo photometry, require the closure of the road section to be captured. This is usually not possible for public roads,

---

[8]Rolling due to uneven roads.

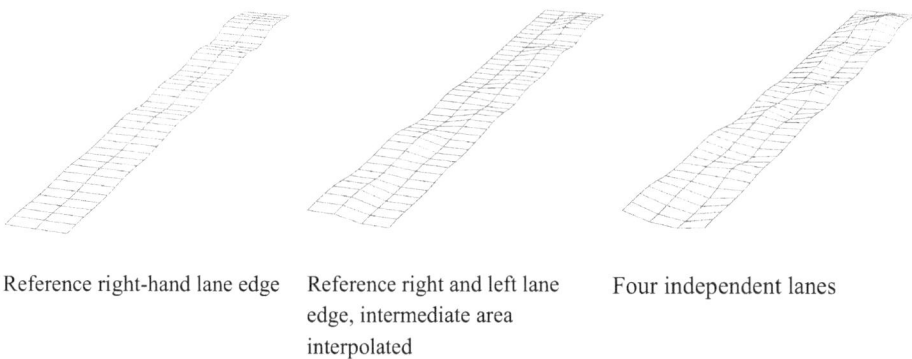

Reference right-hand lane edge   Reference right and left lane   Four independent lanes
                                    edge, intermediate area
                                    interpolated

Fig. 12.16 Road profiles with one, two or four measuring lanes per lane width

so this method can only be used for own internal test tracks. Weather and lighting requirements further restrict this process.

12.5.2 Topology of the Road

In this context, the term topology is used to understand the macroscopic course of the road in the landscape. Here, it has to be differentiated whether it is a two-dimensional $P(x, y)$ or a three-dimensional representation $P(x, y, z)$. Depending on the course of the road, this difference can be significant, because on uphill or downhill roads, the vehicle will slow down or accelerate with the same engine power. Figure 12.17 illustrates that with the example of a section of the Nordschleife of the Nürburgring. Between the start and the Fuchsröhre there is a difference in altitude of about 170 m.

In the case of a comparison between measurement and calculation in which the calculated vehicle is to maintain the speed profile of the measured vehicle, additional braking or acceleration would be required for the flat representation, but that does not correspond to the measurement data. The same applies to the lateral inclination of the track, which occurs mainly on racing and test tracks—extreme on a steep turn. The speeds driven there cannot usually be converted on a flat road surface.

For a complete description of the macroscopic course of the road, the information of the position of the center of the road $P(x, y, z)$, the width of the carriageway $b(x, y)$, their transverse inclination $\varphi\,(x, y)$, if any, and the coefficient of friction prevailing there $\mu(x, y)$ must be known. Between these known points then interpolation must be carried out. The representation in Cartesian coordinates is initially useful, since the route data can be obtained, for example, from GPS data. The required resolution of the position data depends on the application of the road data. For the vehicle dynamics simulation, a relatively coarse scanning in the meter range is sufficient for the x-y-level and in the centimeter range for the z-direction, for the ride comfort and load data prediction, the roughness of the road is

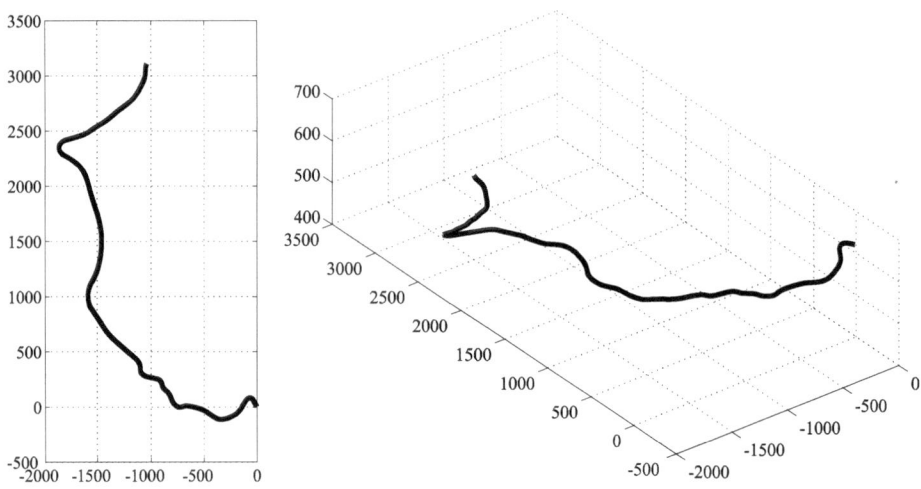

Fig. 12.17   Road course in a two- and three-dimensional representation

required and the road information must be available for the x-y-centimeter level and for the z-direction in the millimeter range.

The example road from Fig. 12.17 lies in a cuboid of approximately $3200\ \mathrm{m} \times 1500\ \mathrm{m} \times 170\ \mathrm{m} = 0.8\ \mathrm{km}^3$. If the street is dissolved in x-y-direction at a distance of 1 m and in the z-direction of 0.1 m, there are 8,160,000,000 possible points between which the tire model must calculate the current road information. If this search is programmed unfavorably, the total of all possible points for each of the four tires is searched for each time step. Considering that the four tires are in close proximity to each other, it is advantageous to search only the first calculation in the entire point cloud and the other three tires, search only in the vicinity of the first tire. If one remembers the last position, the next one, depending on the vehicle speed, will also not be far away, so that a faster search is also possible.

Procedures based on the path coordinate deal with the storage of road information much more economically. Here, the starting point is described at a fixed distance from the course of the road by means of curvature angles and the indication of the road width and the height information over the road width. Since 2008, an open interface definition for the description of road surfaces using this method has been the OpenCRG format (*CRG = Curved Regular Grid*). This format was developed by a VDA[9] working group under the auspices of Daimler AG [GiAR05]. Its goal is to provide every simulation software and every tire model with a transparent, space-saving and fast interface. The various tire models often offer their own road formats and search algorithms, some with significant differences in performance.

---

[9]German Association of the Automotive Industry.

In [WiNe12] a search algorithm is presented, which is based on a representation of the street course in the form of NURBS and should enable a fast discovery of the contact point.

Depending on the measuring method and resolution, very large amounts of data are generated, which usually cannot be processed in this way, but do not have to be processed. For vehicle dynamics simulations, a more macroscopic description of the road course is sufficient, since most tire models only consider flat roads in this area. A trajectory for the driver model is sufficient for the usually flat vehicle dynamics courses on the test tracks. If real racetracks, such as the Nürburgring, are to be simulated, the road width, curbs, curve inclinations, gradients and slopes must also be mapped in order to be able to calculate realistic lap times.

For ride comfort or load data prediction, however, a finer resolution corresponding to the tire model must be aimed at. In [ScAm01], various possibilities are discussed that lead to lossless or lossy data reduction. Particularly over longer distances, very large amounts of data can quickly be generated, which can lead to a noticeable network load and a delay, especially during copying processes (see Sect. 6.3).

12.5.3 Single Events

Single obstacles represent so-called deterministic events, which means that they are clearly defined and lead to a specific excitation. On the one hand, they simulate real roadway situations by synthesizing expansion joints of bridges, lowered or raised manhole covers, level crossings or potholes. On the other hand, in theory, an impact excites all frequencies simultaneously so that the vehicle or component is also excited at its natural frequency.

In driving tests these obstacles are often used for the subjective assessment of the vehicle, as the acoustic impression is dominant here. With a simulation in the time domain, only the effective accelerations of the oscillation can be determined. Validations require measurements which most favorably record accelerations on the wheel carriers, damper domes and seat consoles. That is the best way to detect insufficient parameterization or modeling. Even if the wheel is not independent of the body's movement in the low frequency range, an initial estimate can be made as to whether the differences originate from the tire model or the rest of the chassis.

As already mentioned in the topic Validation (Sect. 3.8.2), it is advisable to measure the maneuver several times and with a high sampling rate. It has a significant influence on the measured acceleration amplitude whether the wheel is moving straight up or down at the beginning of the obstacle crossing. The set of curves generated in this way gives an impression of whether, for example, changes in the chassis tuning can be resolved in relation to the bandwidth of the acceleration amplitudes or not.

Many tire models contain ready-made road models for standard obstacles, which are easy to parameterize, but there is always the possibility to design an obstacle freely.

Often, these individual obstacles are placed on an ideally smooth and even road surface. This facilitates the analysis of system behavior based on this suggestion. However, when

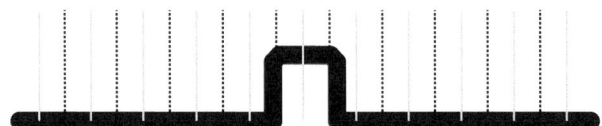

Fig. 12.18 Time-dependent sampling of a single obstacle

mapping real obstacles, it must be taken into account that they may be located on a rough or uneven road surface, so that the influence of the road can also be seen in the measured signals.

Since real single obstacles usually have bevels, the digital versions should also not have ideally sharp edges. The scanning by the tire model is also facilitated in that way, as the already unsteady transition is defused. First and foremost, the scanning capability of the tire model plays the decisive role in determining up to which magnitude contour changes can be detected. Imagine a simple tire model that, every millisecond, provides the information about the current z-coordinate of the street. At a vehicle speed of 30 km/h, which corresponds to approx. 8.3 m/s, the following values are calculated in x-direction every 8.3 mm and a new z-coordinate is determined. Depending on its position on the road, a 10 mm wide and high bar would therefore be covered by the tire at least once (Fig. 12.18, solid line) and at most twice (Fig. 12.18, dotted line). If the strip has a 45° bevel, the determined strip height varies between 9 mm and 10 mm. One sees that here a finer scanning is necessary or in the reverse conclusion that one may only use longer obstacles for this tire model.

With more complex tire models, the number of scanning points above the rolling circumference is often adjustable within limits. The more sampling points are selected, the longer the required computing time. For that reason, this parameter, as far as it is known at all, is often set to smaller values that are just acceptable for the crossing of a stochastic road profile. If single obstacles are calculated, this parameter should be critically checked again. Assuming a rolling circumference of 2 m, 100 sampling points result in a resolution of 20 mm—independent of vehicle speed, since the sampling rate is irrelevant here (Fig. 12.19). The 10 mm wide strip would only be hit by chance. A significant increase of the scanning points is therefore necessary. The resulting increase in computing time is usually not so significant in the short maneuvers.

Since the identification of the tire's model parameters for more complex models is usually based on measurements on test rigs with impact bars, it should be possible to assume that the single obstacle crossing can be mapped relatively well. However, since the previously described applies to the parameterization of the tire model, a critical look at the identification of the tire parameters is once again advisable.

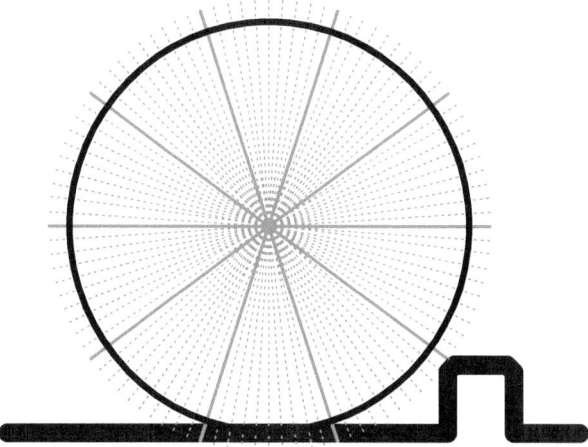

Fig. 12.19 Path dependent scanning of a single obstacle

Fig. 12.20 Periodic excitation

12.5.4 Periodic Excitation

The periodic road excitations also belong to the deterministic excitations. They can be divided into two classes. On the one hand, there are the synthetic excitations. They serve less to map real road profiles than to analyze vehicle behavior. This class is represented by sinusoidal excitation, with which a discrete frequency excitation can take place depending on the vehicle speed and the wheelbase, such as is not to be found in this pure form on real roads. But of course many other synthetic profile shapes can also be found.

The other class depicts real road profiles, which are created, for example, by the use of slab profiles. The same length of the slab profiles results in a periodic effect at constant travel speed. The most prominent example is certainly the Freeway Hop in Los Angeles, where the sagging concrete slabs lead to strong vibration excitation (Fig. 12.20).

12.5.5 Stochastic Excitations

Stochastic excitations are all road profiles that cannot be described with the help of analytical functions, but only with statistical methods. In general, it can be assumed that the roughness profile of a road surface can only be described over its spectrum due to the asphalt or concrete surface.

Individual classes of roads can be grouped together on the basis of this distribution. Thus, city streets differ from country roads or motorways. Even within these classes, distinctions can still be made in the quality of their surface. The underlying amplitude profile can then be converted into a concrete road profile using spectral analysis methods, so that, for example, a synthetic equivalent of a bad country road can be obtained. With these profiles it is possible to carry out basic tests on the vehicle model without having to resort to measurement data from real road profiles. A comparison with vehicle measurements is of course not possible. For this purpose, real street profiles must have been measured and digitized.

References

[Ammo95]   AMMON, D. ET AL: Effiziente Simulation der Gesamtsystemdynamik Reifen-Achse-Fahrwerk, in *Reifen-Fahrwerk-Fahrbahn*, VDI-Berichte 1224, 293-308, Hannover, 1995

[BuMo09]   BURG, H. AND MOSER, A. (ED.): *Handbuch Verkehrsunfallrekonstruktion*, Vieweg +Teubner, Berlin, 2009

[Burg13]   BURGER, M. ET AL: Integration eines detaillierten, flexiblen Reifenmodells in den Fraunhofer Fahrsimulator, in *Reifen-Fahrwerk-Fahrbahn*, VDI-Berichte 2211, Hannover, 167-182, 2013

[Dudz05]   DUDZIŃSKI, P.: *Lenksysteme für Nutzfahrzeuge*, Springer, Berlin, 2005

[DuFS69]   DUGOFF, H., FANCHER, P.S. AND SEGEL, L.: *Tire performance characteristics affecting vehicle response to steering and braking control inputs*. Highway Safety Research Institute, University of Michigan Ann Arbor, 1969

[GaBä07]   GALLREIN, A. AND BÄCKER, M.: *CDTire: a tire model for comfort and durability applications*, in [LuPl07], 69-77, 2007

[GaBB14]   GALLREIN, A., BÄCKER, M. AND BURGER, M.: *Tire Modeling from Structural Analysis to Real-Time Applications*, Simpack News, 03/14, 2-5, Gilching, 2014

[GiAR05]   GIMMLER, H., AMMON, D. AND RAUH, J.: Straßenprofile: Mobile Messung, prozessgerechte Datenaufbereitung und vollständige Bewertung bereiten die Basis für eine effektive Simulation, in *Reifen-Fahrwerk-Fahrbahn*, VDI-Berichte 1912, Hannover, 335-352, 2005

[GiHo14]   GIPSER, M. AND HOFMANN, G.: *FTire: High-End Tire Model for Vehicle Simulation in SIMPACK*, Simpack News, 03/14, 10-15, Gilching, 2014

[Gill92]   GILLESPIE, T.D.: *Fundamentals of Vehicle Dynamics*, SAE, Warrendale, 1992

[Gips01]   GIPSER, M.: Reifenmodelle in der Fahrzeugdynamik: eine einfache Formel genügt nicht mehr, auch wenn sie magisch ist, *MKS-Simulation in der Automobilindustrie*, Graz, 2001

[Gips99]   GIPSER, M.: FTire, a new fast tire model for ride comfort simulations, *International ADAMS User's Conference*, Berlin, 1999

[Hane03]   HANEY, P.: *The Racing & High Performance Tire*, SAE & TV Motorsports, Springfield, 2003

[Klaa99]   KLAAS, A ET AL: TIME, Tire Measurements Eine neue Standardprüfprozedur für stationäre Reifen-Seitenkraftmessungen, in *Reifen-Fahrwerk-Fahrbahn*, VDI-Berichte 1494, Hannover, 119-137, 1999

[Kudr01]   KUDRITZKI, D.: Fahrbahnprofile: Kenntnis und Nutzung in der Fahrwerksentwicklung, in *Reifen-Fahrwerk-Fahrbahn*, VDI-Berichte 1632, Hannover, 255-269, 2001

[LMS09]   LMS INTERNATIONAL: *LMS and Fraunhofer LBF announce strategic partnership for tire modeling and simulation*, press release, 17.04.2009, Leuven

[LuPl07]   LUGNER, P. AND PLÖCHL, M. (ED.): *Tire Model Performance Test (TMPT)*, Supplement to the International Journal of Vehicle System Dynamics, Volume 45, Taylor & Francis, 2007

[OeEF99]   OERTEL, CH., EICHLER, M. AND FANDRE, A.: *RMOD-K – Modellsystem zur Simulation des Reifenverhaltens beim Überrollen kurzwelliger Unebenheiten – Version 6.0*, Manual, 1999

[Oert11]   OERTEL, CH.: *RMOD-K Formula Documentation*, University of Applied Sciences Brandenburg, 2011

[Pace06]   PACEJKA, H.B.: *Tire and Vehicle Dynamics*, SAE International, Warrendale, 2006

[ReRL07]   REINALTER, W., RAUH, J. AND LUTZ, A.: *TMPT – conclusions and consequences für the industry from the industry*, in [LuPl07], 217-225, 2007

[ScAm01]   SCHITTENHELM, H. AND AMMON, D.: *Auf dem Weg zu lebensdauer- und komfortrelevanten 3-dimensionalen Fahrstrecken in der Fahrwerkssimulation*, in *Reifen-Fahrwerk-Fahrbahn*, VDI-Berichte 1632, Hannover, 447-464, 2001

[ScBJ07]   SCHMEITZ, A.J.C., BESSELINK, I.J.M. AND JANSEN, S.T.H.: *TNO MF-Swift*, in [LuPl07], 121-137, 2007

[ScFö09]   SCHMID, A. AND FÖRSCHL, S.: *Vom realen zum virtuellen Reifen – Reifenmodellparametrierung*, ATZ 03/09, 188-193, 2009

[Schr02]   SCHRAMM, E.J.: *Reibung von Elastomeren auf rauen Oberflächen und Beschreibung von Nassbremseigenschaften von PKW-Reifen*, Dissertation, University of Regensburg, 2002

[Uffe80]   UFFELMANN, F.: *Berechnung des Lenk- und Bremsverhaltens von Kraftfahrzeugzügen auf rutschiger Fahrbahn*, Dissertation, University of Braunschweig, 1980

[UnZa97]   UNRAU, H.J. AND ZAMOW, J.: *TYDEX-Format Description and Reference Manual Release 1.3, 2nd International Colloquium on Tyre Models for Vehicle Dynamics Analysis*, Berlin, 1997

[Wase07]   WASER, S ET AL: *Evaluierung von Reifen- und Fahrbahnmodellen für die Simulation festigkeitsrelevanter Beanspruchungen von Nutzfahrzeugen*, in *Reifen – Fahrwerk – Fahrbahn*, VDI-Berichte 2014, Hannover, 153-165, 2007

[Wieg74]   WIEGNER, P.: *Über den Einfluß von Blockierverhinderern auf das Fahrverhalten von Personenkraftwagen bei Panikbremsungen*, Dissertation, University of Braunschweig, 1974

[WiNe12]   WIESEBROCK, A. AND NEUBECK, J.: *User Tire Road Model and Road Sensor für Advanced Vehicle Dynamic Applications*, Simpack News 06/2012, 10 – 13, 2012

[Zege98]   ZEGELAAR, P.W.A.: *The dynamic response of tyres to brake torque variations and road unevennesses*, Dissertation, University of Delft, 1998

Whether the performance characteristics of the engine and transmission, regardless of whether it is an internal combustion engine or an electric motor, are taken into account, depends strongly on the longitudinal dynamics issues to be investigated (Fig. 13.1). Many maneuvers require a constant speed, so the power or torque over the engine speed or the gear changes do not matter. If the engine-gear unit is not modeled with elastic bushings to the vehicle body, not even the mass data need to be taken into account, as they are included in the total vehicle weight.

For some questions, however, this simple consideration is not sufficient, so that more detailed approaches will be discussed below. It is solely a matter of providing the drive torque, not of the processes taking place within the engine or gearbox.

13.1 Specification of the Drive Torque

The fastest modeling option is the neglect of the drive train installed in the vehicle and thus the direct specification of the drive torque at the wheels. This can be done over time (Fig. 13.2), over the path or the vehicle position in the form of a characteristic curve. The time response for the build-up and reduction of the drive torque can be integrated into the characteristic curve. Unrealistic gradients and drive torques should be avoided as they will lead to strongly increased slip values on the tire and thus to implausible driving behavior.

A distinction between drive torque and brake torque can be omitted if, as described in Sect. 16.1, a vehicle speed is specified and this is implemented via a suitable torque control at the wheel.

© Springer Fachmedien Wiesbaden GmbH, part of Springer Nature 2021   239
D. Adamski, *Simulation in Chassis Technology*,
https://doi.org/10.1007/978-3-658-30678-6_13

[BMW Group]

Fig. 13.1 All-wheel drive train of a BMW X4 (F26)

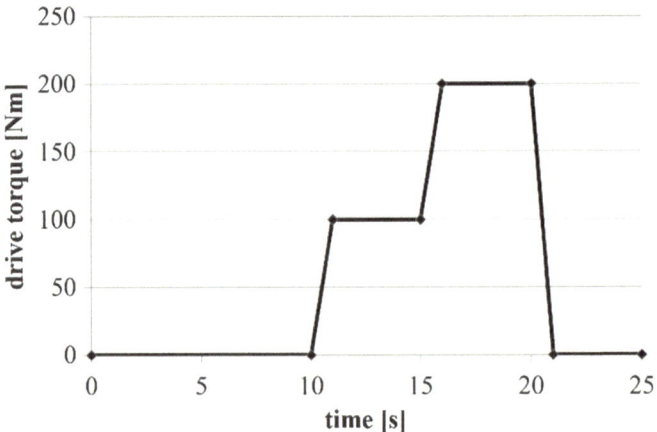

Fig. 13.2 Specification of the drive torque over time

13.2 Engine and Gearbox

13.2.1 Engine Map and Time Response

If the actual performance of an engine is to be represented, the engine torque must be determined via an appropriate characteristic diagram. The three-dimensional engine map (Fig. 13.3) provides the engine torque as a function of the driver's request (accelerator pedal position or engine control request) and the engine speed.

If acceleration processes are to be simulated realistically, the time behavior of the engine-gear unit must also be simulated. This refers above all to the time delay between

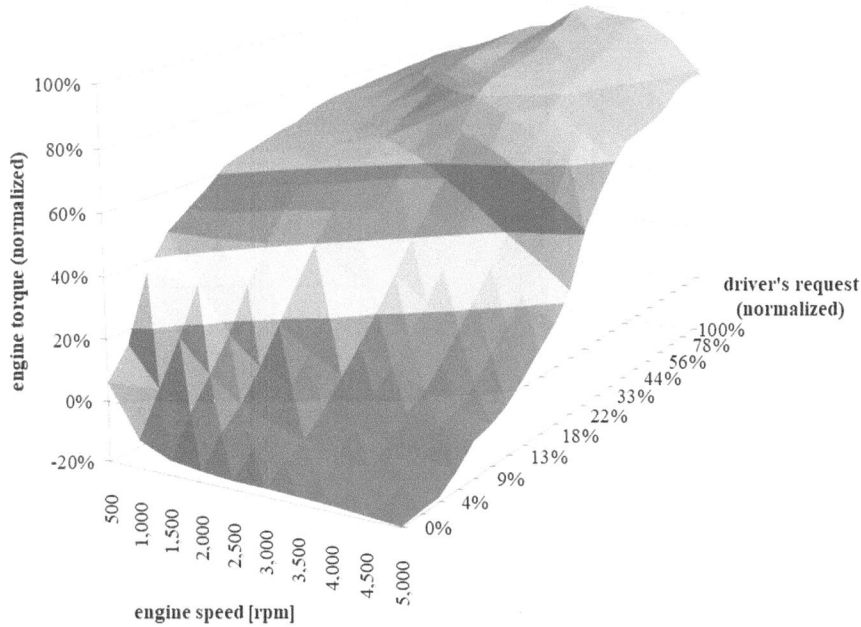

Fig. 13.3 Engine map

the driver's request on the accelerator pedal or throttle valve and the application of the drive torque on the wheel. In addition to the elasticities in the drive train, the mass moments of inertia of all rotating components also play a role here. If the drive train is not to be mapped in detail, appropriate replacement stiffness and moments of inertia must be provided. Alternatively, the time response can be mapped via appropriately parameterized PT_1 elements.

A significant influence on the time response results if the shifting of the driver or the automatic transmission is to be taken into account. This is usually dependent on the engine speed, so it must be calculated, resulting in a much more complex engine model. Now, the performance characteristics of the engine must be stored. If no double-clutch gearbox is installed, a noticeable power interruption occurs during the shifting process, the time behavior of which must also be taken into account. Whether efficiency levels are provided for the clutch and the gearbox, depends on the issue at hand.

13.2.2 Mass Data

The weights of the engine-gear unit vary strongly (approx. 100–500 kg) due to the wide range from three-cylinder petrol engines to ten-cylinder diesel engines, and accordingly also the band of natural frequencies (approx. 4–12 Hz). Of course, only the ready-to-operate engine and the ready-to-operate gear unit with all fluids and attachments are relevant.

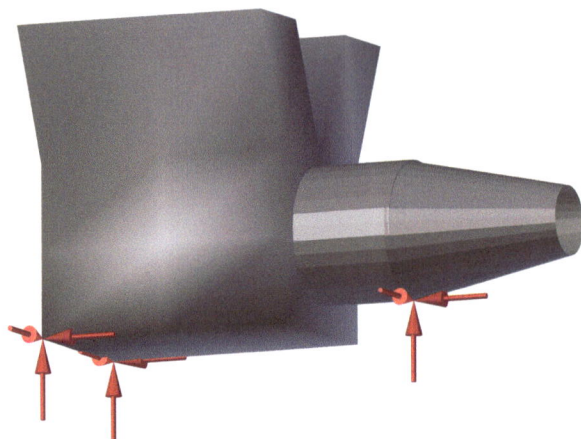

Fig. 13.4 Mounting of the engine-gear unit

Usually, it makes sense to separate the data for the engine and for the transmission, since there is often a choice between a manual and an automatic transmission, depending on the engine. If you want to exchange these variants quickly, the separation of the data of both aggregates offers advantages. The calculation of the common mass data is also saved. The calculation of the new center of gravity and the mass moment of inertia is not rocket science and the theorem of STEINER should be familiar to any engineer, but the software used is capable of this rule as well, so again a possible source of error is excluded. The data is often only available in separate form.

13.2.3 Engine Mounts

The engine-gear unit is mounted to the vehicle body with bushings or torque supports (Fig. 13.4).

The elastic mounting of the engine-gear unit turns the mass block into a vibrating system. The consideration of the bearing arrangement is therefore only meaningful if the natural frequency lies in the relevant frequency range. For vehicle dynamics simulations, the elastic mounting of the engine can often be neglected. In contrast, it is usually necessary for ride comfort and load data prediction.

Conventional elastomeric bearings are implemented exactly like chassis bearings. If hydromounts are used for the bearings, the procedure described in Sect. 8.5 should be used for modeling and parameterization. In the case of switchable or adjustable bearings, it must be taken into account whether, depending on the driving situation, a constant adjustment can be assumed in each case—then the bearing arrangement can also be modeled passively, or whether the adjustability of the bearing arrangement is essential. Then, of course, the active behavior must be considered.

13.3 Axle and Center Differentials

Differentials serve on the one hand to distribute the drive torque, but also to balance the speed between the outer and inner wheel. In all-wheel-drive vehicles, the center differential determines how much of the drive torque is distributed to the front axle or rear axle. Depending on the vehicle, this can be a fixed or driving situation-dependent ratio. If the latter is the case, the criteria for distribution must be considered. If torque vectoring approaches are investigated, wheel individual distribution must be made possible.

In the simplest case, the differential is modeled as a constraint which represents the torque distribution and the initial torques are applied on the wheels of the drive axle(s). If the vibration behavior of the drive train is to be taken into account, the drive shaft should be represented as a mass with a torsional stiffness, especially in rear-wheel drive vehicles. This can be achieved, for example, by an additional torsion spring connected in series. The same applies to the side shafts, which conduct the torque from the axle differential to the wheels. Due to the high torsional stiffness of the shafts (in relation to the elasticity of the bearings), this type of modeling will generally be of large numerical effort, as already explained in Chap. 4.

How detailed the brake system is modeled depends on whether the brake system itself is the object of consideration or whether, for example, the simulation of vehicle dynamics control systems or driver assistance systems require complex modeling (Fig. 14.1). If the primary objective is to reduce or maintain the vehicle speed at a constant level, model approaches such as those described below are generally sufficient. Otherwise, the classic MBS simulation programs are only conditionally suitable for mapping the brake system in detail. For the modeling of a hydraulic brake system, co-simulation with a hydraulic simulation program such as AMESIM or DSHPLUS is necessary. For the modeling of an electromechanical brake, for example, MAPLESIM or MATLAB/SIMULINK could be used. Simulation environments such as MODELICA try to integrate all disciplines into one program regarding the mechatronic character of the system.

Another particular problem is braking when the vehicle is at a standstill, which is dealt with separately in Sect. 14.6. The tire model shall also be taken into account, the reasons why were discussed in Sect. 12.1.3.

14.1 Specification of the Brake Torque

The simplest modeling variant is the neglect of the actual brake system and the direct specification of the brake torque on the wheel. This can be done over time (Fig. 14.2) or over the path or the current position in the form of a characteristic curve. The time response for the production or reduction of the brake torque can be integrated into the characteristic curve. Unrealistic gradients and brake torques should be avoided as they will lead to strongly increased slip values on the tire and thus to implausible driving behavior.

© Springer Fachmedien Wiesbaden GmbH, part of Springer Nature 2021   245
D. Adamski, *Simulation in Chassis Technology*,
https://doi.org/10.1007/978-3-658-30678-6_14

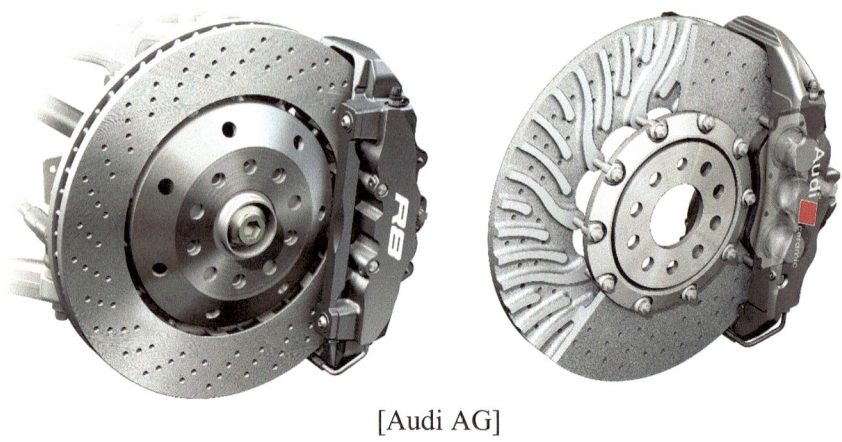

[Audi AG]

Fig. 14.1 Cast iron and ceramic disc of the front axle brake of an Audi R8

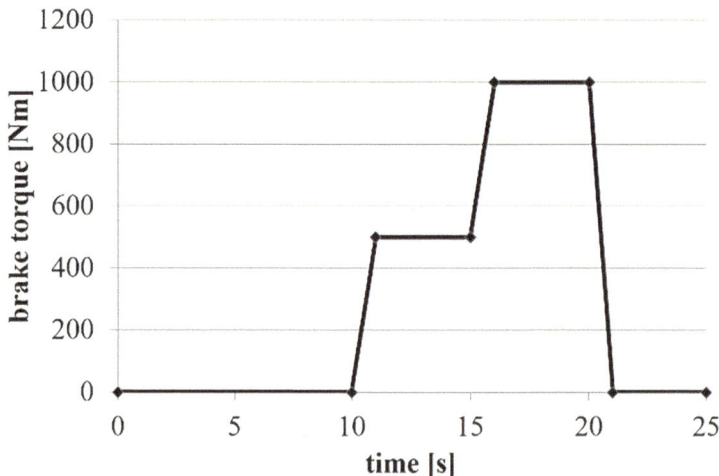

Fig. 14.2 Specification of the brake torque over time

A distinction between drive torque and brake torque can be omitted if, as described in Sect. 16.1, a vehicle speed is specified and this is implemented via a suitable torque control on the wheel.

If you have a déjà vu about the previous chapter, you will see that the difference between a drive torque and a brake torque is only the sign.

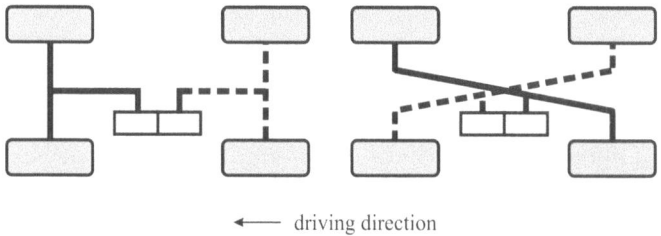

←——— driving direction

Fig. 14.3 Brake circuits according to [DIN74000]

14.2   Brake Circuits

The law stipulates that the brake system consists of two independent brake circuits, so that if one circuit fails, the other circuit is still sufficient to stop the vehicle safely. Whether this circumstance is taken into account in your model depends on whether you want to simulate this failure or not. For passenger cars, the front and rear axle divisions (Fig. 14.3, left) or the diagonal divisions according to [DIN74000] (Fig. 14.3, right) are typically used. In the commercial vehicle sector, there are other variants that take account of the high vehicle weight.

The driver's braking request can be transmitted hydraulically, pneumatically or electrically. In the following, only a hydraulic brake system is described, as it is typically used for passenger cars so far. As soon as the modeling of the hydraulics, i.e. the pipes, valves, pumps and the fluid plays a role as such, the numerical requirements will increase drastically. In [BrBi17], this is illustrated by the integration steps considered necessary, which are shown in Table 14.1.[1]

14.3   Brake Force Proportioning

Depending on the wheelbase and center of gravity in x- and z-direction, each vehicle possesses a so-called ideal brake force proportioning (Fig. 14.4). If that could be implemented in the vehicle, the shortest possible braking distances would result. The related brake forces each represent the relationship between the deductible brake force at the front axle (F_{Bv}) and the rear axle (F_{Bh}) to the weight force of the vehicle (F_G).

In reality, however, the installed brake force proportioning is decisive, which usually results from the ratios of the piston cross-section areas of the wheel brakes of the front and rear axles. If you start from this fixed ratio, you can apply the brake torque to the front and rear axle wheels accordingly. Today's new vehicles, at least in Europe, have an anti-lock

---

[1]If you have skipped Chap. 4 (numerical analysis), maybe you should have a look at it again now.

Table 14.1 Typical integration step sizes for hydraulic simulations

Simulation object	Typical integration step size
Vehicle with wheel suspension	1 ms
Tires	1 ms
Hydraulic lines	100 µs
Hydraulic valves	10 µs
Brake booster with master brake cylinder	50 µs
Brake	200 µs

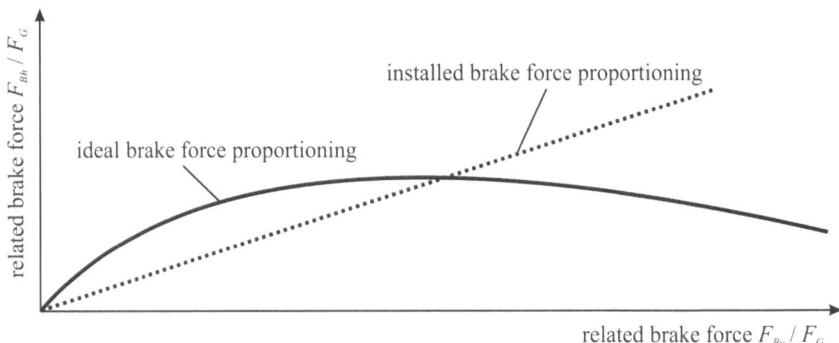

Fig. 14.4 Ideal and installed brake force proportioning

brake system (ABS) that overlays this fixed ratio with an electronic brake force proportioning almost everywhere. Corresponding control strategies can be found, for example, in [Ammo97] or [Reif10].

14.4   Functional Chain from Driver to Wheel Brake

How detailed the functional chain between the foot force of the driver and the clamping force at the wheel brake is described, depends, as usual, on the question posed. It will often be sufficient to apply the brake torque weighted according to the brake force proportioning directly to the wheels during braking. A description of a detailed modeling of the brake hydraulics, i.e. the lines, pumps or valves, as used for the simulation of vehicle dynamics control systems such as the Anti-lock Brake System (ABS), the Traction Control System (TCS) or the Electronic Stability Control (ESC) is necessary, will be omitted at this point.

If, however, realistic actuating forces are to be determined, the brake pedal with its transmission of the foot force, the supporting force of the brake force amplification and the master brake cylinder with the conversion of the actuating force into a pressure must also be taken into account. The resulting hydrostatic pressure now acts as a clamping force on the wheel brake described below via the piston of the wheel cylinder (Fig. 14.5). The

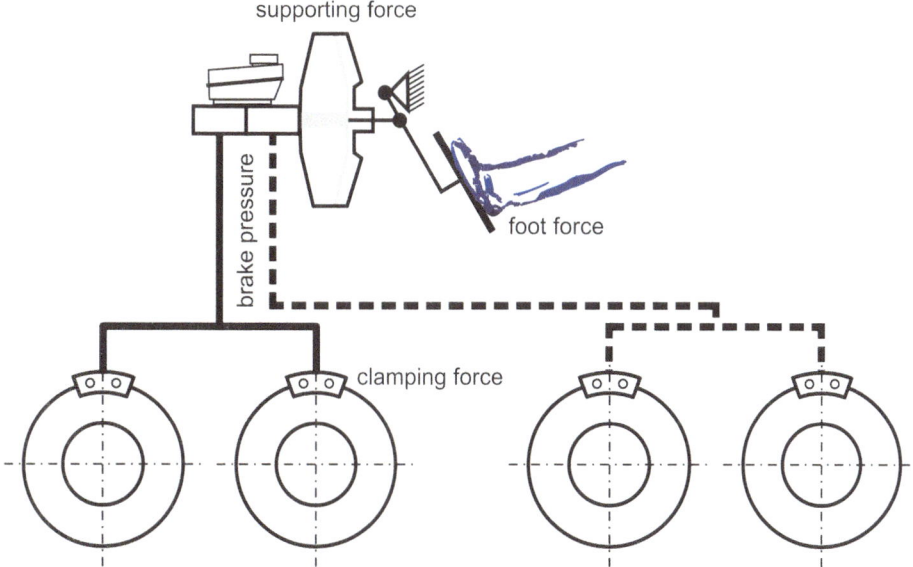

Fig. 14.5 Functional chain in the brake system

driver's braking request is often modeled in a standardized way, i.e. the brakes between 0 and 100% and the pedal force or brake pressure is thus scaled. This corresponds to the idea to represent the brake force (unbraked 0%, partial braking for example 30%, emergency braking 100%).

In the case of classic hydraulic wheel brakes, a distinction must be made between the drum and disc brake designs. However, initially the path from the driver's foot to the wheel brake is identical in both systems. The so-called external ratio, which describes the relationship between the clamping force and the foot force to be applied, is therefore described first. The so-called internal ratio of the brake or also the brake characteristic value is then used in accordance with the design.

The following calculations do not take friction or efficiency into account. Where such data are available and need to be taken into account, they shall be completed in the appropriate place.

Due to the lever ratio of the brake pedal (Fig. 14.6), the following moment equilibrium occurs:

$$F_{HBZ\_Ped} \cdot a = F_{Ped} \cdot b, \tag{14.1}$$

so that the force F_{HBZ_Ped} results from the pedal force F_{Ped} at the master brake cylinder.

Fig. 14.6 Brake pedal ratio

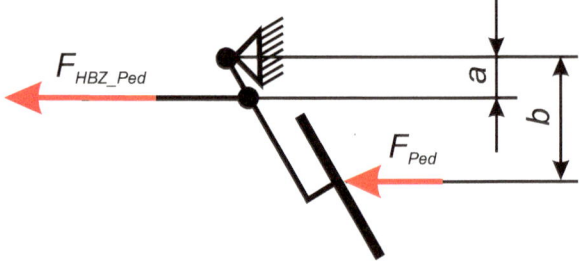

$$F_{HBZ\_Ped} = \frac{b}{a} \cdot F_{Ped} \qquad (14.2)$$

Figure 14.7 shows the ratio at the pedal with five (dotted line). The supporting force of the brake booster is added to this proportion, so that the total force acting on the master brake cylinder is F_{HBZ} consists of the sum of these two parts.

$$F_{HBZ} = F_{HBZ\_Ped} + F_{HBZ\_Boost} \qquad (14.3)$$

However, the brake force support must be limited upwards at the knee point. This is possible, for example, with the following function:

$$F_{HBZ\_Boost} = \min\left(i_V \cdot F_{HBZ\_Ped}; F_{AP}\right) = \min\left(i_V \cdot \frac{b}{a} F_{Ped}; F_{AP}\right) \qquad (14.4)$$

i_V Amplification factor of the brake booster
F_{AP}
   Maximum force of the brake booster at the knee point

The sum of both forces acts on the piston cross section A_{HBZ} of the master brake cylinder, which then generates a pressure p_{Brems} for the two brake circuits (Fig. 14.8). Depending on the design, spring elements in the brake booster and master brake cylinder must also be taken into account. Usually, they are pre-loaded and additionally apply a displacement-dependent force. For the sake of clarity, they are not listed here.

$$p_{Brems} = \frac{F_{HBZ}}{A_{HBZ}} \qquad (14.5)$$

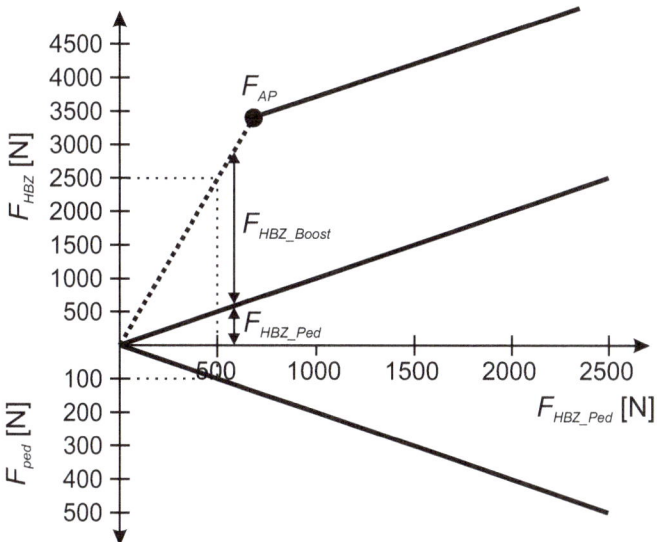

Fig. 14.7 Amplification of pedal force, according to [BrBi17]

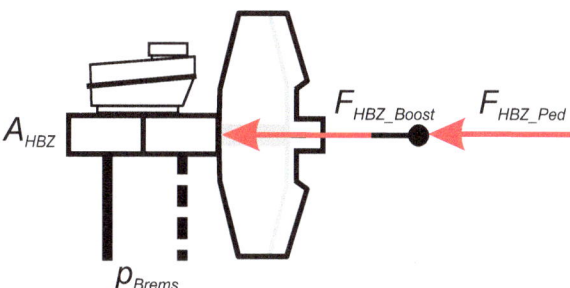

Fig. 14.8 Generating the brake pressure

If the pressure loss and the time behavior is neglected due to the brake lines, this pressure acts on the piston of the cylinders in the wheel brakes A_{RZ} and generates the clamping force F_{SP} there.

$$F_{SP} = p_{Brems} \cdot A_{RZ},\qquad(14.6)$$

so that finally the external ratio $i_{\ddot{a}}$ results in the following:

$$i_{\ddot{a}} = \frac{F_{Sp}}{F_{Ped}} = \frac{A_{RZ}}{A_{HBZ}} \cdot \left(\frac{b}{a} + \frac{F_{HBZ_{Boost}}}{F_{Ped}}\right).\qquad(14.7)$$

For simplicity's sake, if we assume that the piston diameter of the wheel cylinder is twice as large as the piston diameter in the master brake cylinder, this results in a pre-factor of four. If we continue to assume from Fig. 14.7 that the lever ratio on the pedal is five, without the brake force increase the foot force would already be increased by a factor of 20. If an additional force comes from the brake booster which, for example, amplifies the foot force already amplified by the lever transmission five times, the clamping force at the wheel brake in this example is 100 times higher than the applied foot force before reaching the knee point. Depending on the design of the brake, this clamping force is applied either radially (drum brake) or axially (disc brake).

14.5 Brake Torque at the Wheel Brake

14.5.1 Drum Brake

The first distinction to be made between drum brakes is their design. Depending on the design, one leading and one trailing or two leading brake shoes would have to be modeled. Friction values that change due to contamination or temperature differences create a non-linear relationship between the clamping force and the brake force.

Figure 14.9 shows a simple drum brake. As an example, the resulting brake torque M_B of a brake shoe is to be calculated here. Assuming COULOMB's friction between the brake force F_B and the normal force F_N exists the following relation

$$F_B = \mu_B \cdot F_N \tag{14.8}$$

If, as in [BrBi17], a moment equilibrium around the lower pivot point A is set to

$$F_B \cdot r_{\mathit{eff}} = F_N \cdot a - F_{Sp} \cdot 2a = \frac{F_B}{\mu_B} \cdot a - F_{Sp} \cdot 2a \tag{14.9}$$

the result for the brake force F_B is

$$F_B = F_{Sp} \cdot \frac{2a\,\mu_B}{a - \mu_B r_{\mathit{eff}}} \tag{14.10}$$

and thus for the brake torque M_B

$$M_B = F_{Sp} \cdot \frac{2a\,\mu_B r_{\mathit{eff}}}{a - \mu_B r_{\mathit{eff}}} \tag{14.11}$$

The internal ratio, also known as the C*, is the relationship between the clamping force and the brake force.

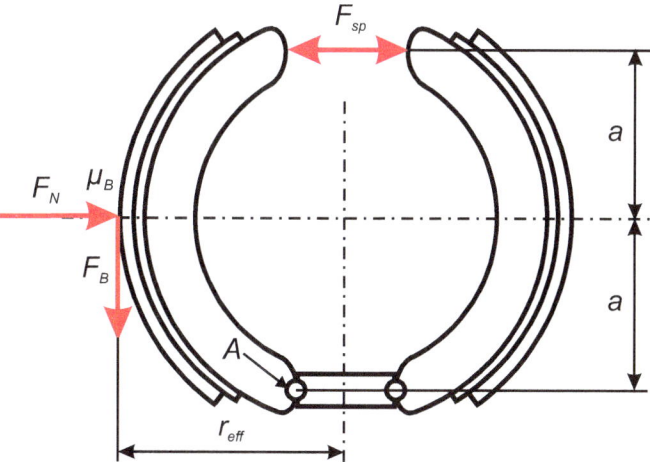

Fig. 14.9 Simple drum brake

$$C^* = \frac{F_B}{F_{Sp}} = \frac{2a\,\mu_B}{a - \mu_B r_{eff}} \tag{14.12}$$

The total ratio between the pedal force and the brake force of the drum brake is therefore made up of the Eqs. (14.7) and (14.12).

14.5.2 Disc Brake

Modeling a disc brake (Fig. 14.10) is relatively simple in most applications, as there is a linear relationship between the clamping force and the brake force generated above the coefficient of friction.

In the simplest case, the resulting brake torque is modeled with COULOMB's friction.

$$M_B = 2 \cdot \mu_B \cdot F_{SP} \cdot r_{eff} \tag{14.13}$$

The internal ratio or the brake sensitivity C^* simplifies to

$$C^* = \frac{F_B}{F_{Sp}} = 2\,\mu_B \tag{14.14}$$

The total ratio between the pedal force and the brake force of the disc brake is therefore made up of the Eqs. (14.7) and (14.14).

Fig. 14.10 Disc brake

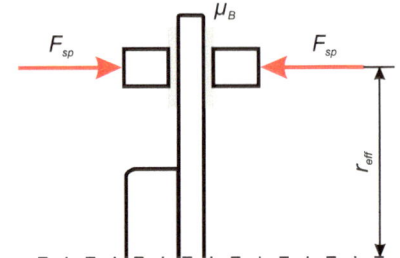

14.6 Braking to a Standstill

If the brake torque on the wheel is modeled as in the previous sections, this is well suited to maintain a speed (steady-state braking) or to decelerate from a higher speed to a lower speed sufficiently greater than zero (deceleration braking). This procedure is not suitable if you have to brake to a standstill. At least not without further aids.

What is the problem? In the simplest case, a constant brake torque is applied. The vehicle is initially decelerated, reaches zero speed and is now supposed to stop, just as a real vehicle would. In reality, the brake torque now becomes a drive torque and the vehicle is accelerated backwards. The real brake torque is produced by a friction force which is opposed to the movement of the wheel and which changes from sliding to stopping in the event of a standstill. In the procedure described above, there is no difference between a drive torque and a brake torque.

Is it not possible to simply set the brake torque to zero as soon as standstill has been reached? In only a few cases this will lead to the vehicle really stopping, because the pitching movement that the vehicle performs due to the omitted deceleration or an inclined road surface would lead to a renewed movement.

Some simulation programs allow freezing of degrees of freedom. This means that the degrees of rotational freedom of the wheels must be locked at the moment of standstill. This changes the dimension of the state vector and one changes an essential system property. Here, state events must be defined, which lead to the switching off and later again to the switching on of the degrees of freedom.

Another variant would be the more detailed modeling of the friction contact, so that as long as the brake pressure is applied, the brake actually remains in the static friction during standstill.

The bottom line is that if you have to brake to a standstill, you should first check how the model reacts to it. The combination of tire and brake model can sometimes cause surprises.

14.7   Coefficient of Friction and Temperature Behavior

With a real vehicle it makes a big difference whether you measure the braking distance with a cold, a warm or an overheated brake, for example. For most calculations it will be sufficient to specify the coefficient of friction between pads and disc or between pads and drum as a constant parameter. This parameter can then be assigned to a cold or warm brake according to the question posed. This also applies analogously to a damp or dirty brake.

If, however, the heat balance of the brake has to be examined, the modeling becomes much more complex. The first step is to calculate how much energy is applied to the brake in the form of heat during operation. Since the brake torque is known, this must simply be balanced over time, i.e. there must be an energy store that accumulates the current energy content. On the other hand, there is the energy which is transported out of the brake system by heat dissipation (heat conduction, radiation, convection). Convection depends on the air flow around the brake, which in turn depends on the vehicle speed and the way the air reaches the brake. Here, one should be able to fall back on the corresponding characteristic maps of the flow simulation or the wind tunnel measurements. In addition, it must also be known how the heating of the entire brake system is distributed among the individual components. How do the brake pads and brake fluid heat up? What is the fading behavior?

References

[Ammo97]   AMMON, D.: *Modellbildung und Systementwicklung in der Fahrzeugdynamik*, Teubner, Stuttgart, 1997

[BrBi17]   BREUER, B. AND BILL, K. H. (ED.): *Bremsenhandbuch*, Springer Vieweg, Wiesbaden, 2017

[DIN74000]   DIN 74000: *Hydraulische Bremsanlagen: Zweikreis-Bremsanlagen – Kurzzeichen für die Bremskreisaufteilung*, 1992

[Reif10]   REIFF, K. (ED.): *Bremsen und Bremsregelsysteme*, Vieweg+Teubner, Wiesbaden, 2010

15

15.1 Body in White

For many questions it is sufficient to represent the mass, the moment of inertia and the center of gravity of the body in white by a rigid body (Fig. 15.1). Depending on the required variability, further rigid masses are then connected to this body mass either elastically (engine-gear unit, exhaust system, etc.) or rigidly (passengers, load, etc.).

There are two main reasons for moving away from this simple and quick procedure.

1. The static deformation of the body in relation to the chassis is so large that it can no longer be neglected. This is particularly the case in the commercial vehicle sector. The long, relatively torsional soft frame structures of buses or trucks cannot be reproduced sensibly by rigid bodies.
2. The dynamic vibration behavior of the body and its attachments is essential for the effect to be considered. This occurs above all with ride comfort or load data prediction simulations. For the former, the vibration excitation of the structure by, for example, roadway excitations is of more interest, for the latter, the local deformation of the structure under external forces.

Normally, an OEM has an oscillating FE model of the body, which can be the basis for the integration of the structural behavior. These models have a very large number of degrees of freedom and cannot be used unprocessed in the world of MBS. Most commercial MBS simulation environments offer interfaces to the common FEA programs, which simplify the integration of flexible structures from the FEA world, but some preliminary work has to be done, which is briefly described below.

© Springer Fachmedien Wiesbaden GmbH, part of Springer Nature 2021   257
D. Adamski, *Simulation in Chassis Technology*,
https://doi.org/10.1007/978-3-658-30678-6_15

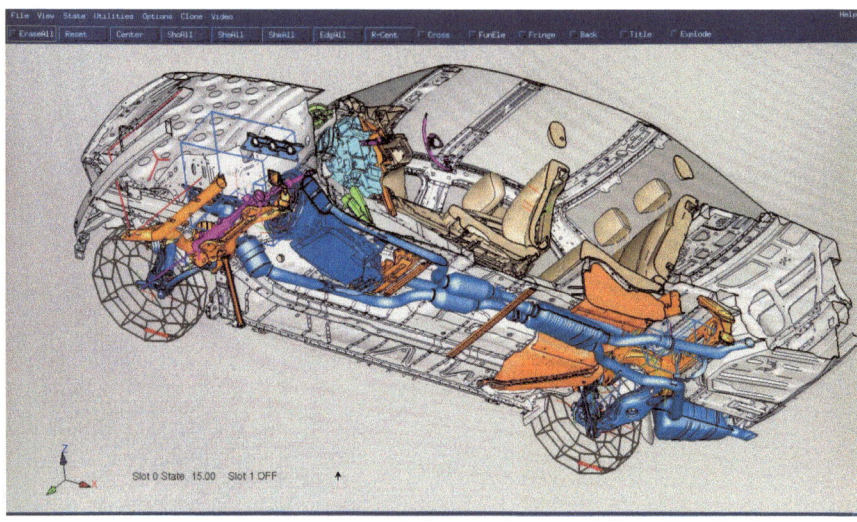

[Daimler AG]

Fig. 15.1 FEA full vehicle model of a Mercedes C-Class (W204)

The MBS and FE models do not usually originate from a single source. Since most MBS engineers only have rudimentary FEA knowledge, cooperation between the two groups is particularly important here.

15.1.1 Preparation of the FE Model

First of all, it must be ensured in the FE model that all connection points of the MBS model are also present. A connection point, for example, is the geometric location of a bearing at which the vehicle body and the suspension are to be connected kinematically. This coordinate must be exactly the same in both models,[1] i.e. there must be a node in the FE model at this point. That will not always be the case. If we stick to the bearing example, in the FE model the mounting for the bearing is modeled in such a way that the bushing can be pressed in accordingly. However, the MBS model does not provide for an extension of the bearing and only knows the kinematic point as a geometric location. An RBE2 or RBE3 element is often used for the connection, depending on how stiff the force is to be introduced into the FE model. This element must then be linked to a sufficient number of

---

[1]The required accuracy depends on the programs you use. Usually, an error message will prevent you from being able to calculate until the coordinates are sufficiently accurate.

neighboring nodes in order to represent the introduction of force as realistically as possible. An RBE2 element represents a rigid connection between a master node and the nodes involved, resulting in local stiffening. With the RBE3 elements, the displacements of the connected nodes are recalculated on the master node, which is less stiff in its effect.

15.1.2 Modal Reduction

Once the FE model has been prepared as described in the previous section, a reduction method is used to reduce it to a size suitable for the MBS program. The number of degrees of freedom in a complete vehicle model is too large by orders of magnitude to be treated with a MBS algorithm. The reduction methods of GUYAN or CRAIG-BAMPTON are sufficiently described in the literature, so that I will only go into it very briefly at this point [GeCa01]. The goal is to drastically reduce the number of nodes to be considered. For this purpose a transformation matrix has to be developed, which transfers the old, too large state vector to the new, reduced state vector.

In short, the reduction methods divide the nodes of the FE model into inner and outer nodes, which are subsequently treated differently. The outer nodes represent the connection points to the MBS model. The masses are concentrated on them and the external forces also apply there. The inner nodes describe the system behavior within the FEA world. The eigenmodes of the system are subdivided into *Constraint Modes* and *Normal Modes*. The *Constraint Modes* are determined by fixing all but one of the outer nodes. Each node is then individually loaded with a unit load and its static deformation is determined. The *Normal Modes* then represent the eigenmodes of the system fixed at the outer nodes. The user can determine how many and which eigenmodes are taken into account. This step requires a lot of experience with the method and the selection of what is relevant for the load case and what is not. You will select the eigenmodes based on their frequencies, starting with the lowest ones.

The choice of the method and the number of modes used in the CRAIG-BAMPTON-methods have a great influence on the deviations of the eigenmodes in the model to the original, but also on the required computing time for the generation of the necessary data and the later calculations [Benz08].

The FEA programs usually offer the user sufficient selection options to determine the number of eigenmodes required, the size and distribution of modal attenuation, etc. The FEA programs can also be used to determine the modal response of the modal damping. As a result, the FEA programs supply files which the MBS programs can then evaluate. Depending on the tools used, the procedure may differ in detail. This is described in [Geig11] for ANSYS and SIMPACK.

15.2   Total Mass

15.2.1 Mass Distribution

As long as you stay in the MBS world, reaching the weight of the vehicle is easy. All modeled components have their mass, their center of gravity and their moment of inertia. If one subtracts the sum of all modeled components from the vehicle mass, then that remains, which must be added to the main body in the simplest case. In this way, it is relatively easy to adjust the mass, the position of the center of gravity and the moment of inertia for the entire vehicle. If you can assign a part of the mass to the unsprung mass, you should of course also place it there.

It becomes more problematic if the body with its attachments is added as a FE model. The FEA colleagues whose FE models are used will also have provided everything they have modeled with the correct masses and stiffness. However, the principle of caution applies here. Especially when you access such a model for the first time, you need to know for which purposes the model should be used and how the target weight was reached. Some modelers also work in the FE model with so-called *lumped masses*, with mass points at which the mass is concentrated. So only because, for example, the engine cannot be seen as a structure in the FEA program, it does not mean that it was not considered as a mass.

Of course, it must first be clarified that no components that are mapped in the MBS model are still present in the FE model. This usually applies to all chassis parts, the engine-gear unit, the drive train and, if applicable, the exhaust system. If you now add the masses from the FE model and the MBS model, there are three possibilities.

1. The sum of all masses gives exactly the target mass, the mass distribution is in the right place and the total moment of inertia is correct. Before you start cheering loudly and slapping on your shoulder, you should calculate again. This case is so unlikely that I will not deal with it further.
2. The sum of all masses is below the target mass. This is the most probable case, because even in the FE model, everything that has not been modeled weighs nothing. Here, a concept must be considered how the total mass (trivial), the center of gravity (almost trivial) and the total moment of inertia (rarely trivial) can be achieved with several compensating masses.
3. The sum of all masses is above the target mass. The most unfavorable variant. Assuming that the target mass is correct, masses must have been overestimated in the MBS or FE model. It does not help, you have to check component by component and compare the masses with the respective construction data or weighed masses. The approach: "Well, then we lower the density for the body a bit, then it will fit again", forbids itself. After all, we want to map the vibration behavior of the body. As you know, there is a connection.

The second case is the one that will be most common. So the question is, how can I best meet the requirement that the total mass, the mass distribution and the inertia properties should be right?

15.2.2 Use of One Correction Mass

Imagine we have a mass difference of 100 kg. Let us start with the approach that often comes to mind first. We add a mass point with any position. Total mass and mass distribution can thus be corrected quickly. We probably would not be able to influence the inertia tensor favorably, because it shows the mass distribution in space and we work with a concentrated mass. But maybe we only want to examine micro shake and only drive straight ahead on an even road, so that neither yawing, pitching nor rolling play a big role. Ready? Unfortunately no, because we still have to fix this mass in the vehicle. How should this attachment look in the best case and to what should the mass be attached? If we manipulate the FE model, a new FE model must be generated for each mass change (e.g. different loading conditions). So we better stick them to the MBS model, that is faster changed. The mass point will probably have to be placed close to the overall center of gravity in order to achieve the correct mass distribution. The connection to the model could be established by a rigid massless connection. We cannot attach a mass point of 100 kg somewhere to the chassis, i.e. to the unsprung mass, so we have to go to the side of the sprung mass. These would be, for example, all connection points of the chassis to the body. At this point we have to ask ourselves again why we want to integrate a FE model of the body. We want to map stiffness behavior, i.e. elasticity, better. It is obvious what will happen if we hang 100 kg from a lever and attach it to a connection point. These *points* are designed to be relatively stiff in order to support the forces and torques from the chassis to the body without large deformation. But this is an *unnatural* load which is not provided for here and which will in any case lead to unintentional deformations and an unbalanced vibration behavior. Well, then we fix the mass at several points so that the load is better distributed. The four damper domes, which are also very stiff because they have to support a large part of the dynamic wheel loads, would be ideal for this. We now have a mass point that hangs on the damper domes with four rigid struts. This is a constraint for the domes since they can no longer change their distance to each other.[2] Something we wanted to achieve by using the flexible body structure. If it does not depend on the overall vibration behavior of the body, but only on the introduction of force from axle components into the body, this error may be acceptable. The results of a ride comfort simulation will be characterized by the behavior of a body that is too stiff. Then we do not use rigid connections, but give them a stiffness similar to that of a car body, then the four points can move relative to each other again. This, however, also applies to the single mass. We

---

[2]In the real vehicle, this would correspond to a diagonal brace, which is also used for stiffening.

would have created a single mass oscillator whose natural frequency we can explicitly read off in the frequency evaluation.

15.2.3 Use of Several Correction Masses

Lever arms described at previous section thus are a problem. If we use four instead of one correction mass, we could place them directly on the damper domes. The lever arm is omitted and it is still sufficiently easy to adjust the mass distribution at least in the x-y-level to be set. This does not apply to the z-coordinate, because the position on the damper dome means that the overall center of gravity will usually be above the target center of gravity. This would only be possible by shifting the correction mass along the z-axis—with that, however, another lever arm is formed. Here it has to be decided which error is more important for the load cases under consideration—the wrong center of gravity height (and thus e.g. a larger lever arm between center of gravity and roll axis) or the influence on the point where force is applied to the body.

Since our mass points each have their own inertia tensor, we get more possibilities to reach the desired total inertia tensor. Of course, the four masses still have a retroactive effect on the body. But since each of us in our example will weigh only 20–30 kg, the error is usually smaller than before.

15.2.4 Conclusion

It can be seen that the best solution is to map the weight of the modeled components as accurately as possible. This also means that for a chassis strut, for example, not only the weight and the mass properties of the strut have to be taken into account, but also those of the joints and bearings, of screws, cables, shielding and wind deflectors and of all other existing components which are not modeled in the MBS world. The lighter the correction masses are, the less interference they have. They are and remain only a makeshift solution, the retroactive effect of which must always be kept in mind.

15.3   Aerodynamics

Aerodynamics is a topic in its own right, which is also investigated mathematically with the aid of CFD simulations both for the flow around and for the flow through. The shape of the outer skin, the engine compartment or the interior is essential for this. The required computing time of a CFD application is clearly too high for vehicle dynamics investigations and the vehicle parts flowed around and through must be available with their surfaces at the time of the calculation. In the context of the simulation tasks discussed so far, this information is generally not known or is not taken into account in the programs used.

Nevertheless, the influence of aerodynamics on driving behavior, especially at higher speeds, is essential and must not be ignored. Their effect is taken into account by using applied forces, which then represent the resistance, lateral and buoyancy or downforce forces. This way, driving resistances, crosswind influence and axle load changes can be viewed as a function of wind speed and wind direction. The characteristic maps and values required for this are then derived from CFD calculations or wind tunnel measurements.

A comprehensive overview of the topic of aerodynamics in relation to vehicles can be found in [Schü13].

15.3.1 Wind Resistance

Driving resistances are the forces which are opposite to the direction of travel. Every cyclist knows the feeling, to *seemingly* drive against the wind all the time. The best known parameter is the c_W-value, which is the drag coefficient on everyone's lips. However, it is only a part of the truth, because only in connection with the corresponding frontal surface A you can compare two vehicles in terms of their drag. The c_W-value, however, only describes the inflow of the vehicle at $0°$, i.e. in x-direction, so that in practice the tangential coefficients $c_T(\tau)$ which were determined under lateral inflow. At a given air density ρ the amount of aerodynamic drag F_{Lx} is calculated with stationary air as a function of vehicle speed v.

$$F_{Lx} = \frac{\rho}{2} \cdot c_W \cdot A \cdot v^2 \tag{15.1}$$

The wind resistance limits the maximum speed of the vehicle, since at some point a balance is reached between the driving force attainable by the engine power and the resistance force. This way, driving performance and fuel consumption can be estimated. If these variables do not play a role because, for example, constant velocities are calculated, the consideration of wind resistance can also be neglected.

If the vehicle is equipped with a level control system which lowers the vehicle level at higher driving speeds, the front end area and the c_W-value of the vehicle, resulting in lower resistance.

15.3.2 Crosswind

As soon as the vehicle side-streamed, several effects occur simultaneously. Of course, the crosswind also has a longitudinal component (except with a flow of less than $90°$) and thus contributes to the driving resistance. However, what is essential is the resulting lateral force, which generates a yaw moment in the vehicle due to the lever arm between the center of gravity of the vehicle and the point at which the force is applied (pressure point) and thus forces the vehicle out of its target track. Especially when the side angle is jerky, as can be the case with valley bridges or forest clearings, the steering response demands the driver's full attention.

The tangential coefficients discussed above are applied to the same surface A as with the longitudinal approach flow. Thus, as one might assume, the lateral projection surface is not determined as a function of the wind angle τ is used.

$$F_{Ly} = \frac{\rho}{2} \cdot c_T(\tau) \cdot A \cdot v_R^2(\tau) \tag{15.2}$$

The speed v_R is no longer the driving speed, but the resulting inflow velocity, which can be calculated as the amount from the superposition of the wind velocity v_W and the vehicle speed v results in

$$v_R = \sqrt{v^2 + v_W^2 + 2 \cdot v \cdot v_W \cdot \cos \tau} \tag{15.3}$$

When a crosswind occurs, one assumes a force with a constant direction and constant magnitude. On some test tracks there are wind turbines that, with constant wind speed, can direct the wind flow sideways at the vehicle. Here, you can measure the course deviation of the vehicle with the steering wheel held or, when trying to follow the given course, view the steering effort of the driver. Since the pressure point is usually located between the front axle and the center of gravity of the vehicle, in this case it is necessary to steer against the wind direction.

On the other hand, the cause can be natural crosswind, which can then have a stochastic effect on the vehicle with different directions and amounts. This scenario corresponds to the customer situation and is suitable, for example, for testing an algorithm for crosswind compensation with a controlled steering intervention of an EPS steering system.

15.3.3 Buoyancy

Another effect caused by the flow around the outer skin of the vehicle is the lift or downforce of the vehicle. While in an aircraft, in order to be able to take off, one wants to achieve lift through the shape of the wings, the vehicle should of course remain on the ground. If you consider that small sport planes can take off from the ground already at approx. 100 km/h, you should also consider this effect with the vehicle during maneuvers with higher speed. Otherwise, the calculation of most vehicles would suggest that the driving behavior is too stable.

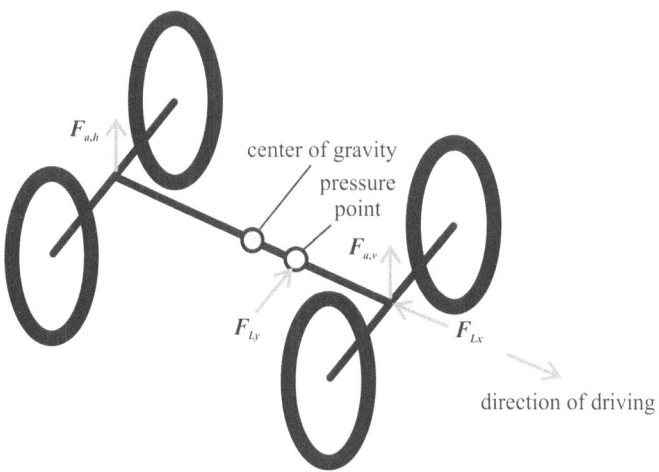

Fig. 15.2 Effect of aerodynamic forces

The shape of a vehicle provides lift without further aerodynamic measures, which ultimately contributes to a reduction in wheel loads. You have to distinguish between the lift coefficient of the front axle $c_{a,v}$ and that of the rear axle $c_{a,h}$. The axle-wise reduction of the load has a direct influence on the self-steering behavior of the vehicle. The buoyancy forces can be regarded as applied forces on each axis.

$$F_{A,v} = \frac{\rho}{2} \cdot c_{a,v} \cdot A \cdot v^2 \tag{15.4}$$

$$F_{A,h} = \frac{\rho}{2} \cdot c_{a,h} \cdot A \cdot v^2 \tag{15.5}$$

If the front axle is relieved more than the rear axle, there will be understeering, in the opposite case oversteering. The effect of all aerodynamic forces described can be seen in Fig. 15.2.

With the help of aerodynamic measures such as a spoiler, a downforce can be generated or the lift can be reduced. In motor racing, large downforce forces are used, so that up to 2–3 times the vehicle's weight can be achieved [Tres12].

References

[Benz08]   BENZ, R.: *Fahrzeugsimulation zur Zuverlässigkeitsabsicherung von karosseriefesten Kfz-Komponenten*, Dissertation, University of Karlsruhe, 2008

[GeCa01]   GÉRADIN, M. AND CARDONA, A.: *Flexible Multibody Dynamics*, John Wiley & Sons, Chichester, 2001

[Geig11]   GEIGER, N. ET AL: New methods for chassis modeling of commercial vehicles an application for dynamics and durability simulation in *2nd International Munich Chassis Symposium*, 319-338, Munich, 2011

[Schü13]   SCHÜTZ, TH. (ED.): *Hucho – Aerodynamik des Automobils*, Springer Vieweg, Wiesbaden, 2013

[Tres12]   TRZESNIOWSKI, M.: *Rennwagentechnik*, Springer Vieweg, Wiesbaden, 2012

The Simulated Driver

16

The driver's interfaces to the chassis-specific vehicle functions are the steering wheel, the accelerator and brake pedals and, depending on the type of transmission, the clutch pedal and gear selector lever (Fig. 16.1). If driver assistance systems or vehicle dynamics control systems are to be mapped, the number of control elements increases accordingly. Whether these interfaces can be found as components in the model or whether the driver's wishes are transferred in another way depends on the necessary modeling depth. For example, the steering request can be applied by a steering angle or a steering torque on the steering wheel, on the pinion or directly on the rack (as displacement or force) (see Chap. 11).

First, we have to distinguish between open-loop and closed-loop maneuvers, depending on whether a feedback is necessary or not.

In open-loop maneuvers, it is therefore not checked whether the input parameters lead to the desired result. No sensors or controllers are required here (the driver does not intervene in the driving test, in some cases the specifications are generated, for example, by a steering machine). It is sufficient if a manipulated variable has been defined over time or path.

In contrast, closed-loop maneuvers check whether the target value (e.g. the vehicle's speed) has been reached. In the event of deviations, readjustment is made with the aid of a manipulated variable (e.g. accelerating or braking). This requires a sensor system and a controller to reduce the actual deviation. In real driving tests this is usually the driver in personal union. However, the autonomous driving of the vehicle must then actually be implemented in a control unit with a sensor system (camera, radar, etc.) and a control algorithm. In [Mayr01] it is described in detail what has to be considered for the longitudinal and lateral control of driver assistance systems.

© Springer Fachmedien Wiesbaden GmbH, part of Springer Nature 2021
D. Adamski, *Simulation in Chassis Technology*,
https://doi.org/10.1007/978-3-658-30678-6_16

[BMW Group]

Fig. 16.1 Driver interface steering wheel

16.1 Speed Control

16.1.1 Initial Value

The principle of setting up the equations of motion as first-order differential equations using the numerical solution, as described in Sect. 4.2, generates a so-called initial value problem. A value must therefore be specified for each degree of freedom and its derivative so that the solution procedure can start. You can take advantage of this and use it for the x-position and the x-speed of the vehicle (of course also for all other degrees of freedom) can be specified. These are then the initial values of the simulation, i.e. the vehicle starts at the specified position with the specified speed. Strictly speaking, this approach has no place at this point. But experience shows that some people fall for this fact and assume that the vehicle will maintain this speed in the future. Of course, this is by no means the case—as long as no suitable controller is available. So, if it is observed that the vehicle slows down steadily from the desired speed, this *false* approach could be the cause. Because only for the first integration step is this speed (and the position) decisive. Then the modeled physics should work so much that the vehicle slows down.[1] Of course, it is not enough just to give the initial speed to the body of the vehicle; all rotating bodies (wheels and drive train) must also have the correct starting speed. Otherwise, you put a vehicle on the road that already has a speed, but the wheels and the drive train are stationary and have to be accelerated only because of the road contact. Not a very realistic idea. In order to not have to preset all

---

[1]Unless you are driving downhill.

speeds manually, it is recommended to predetermine a parameter for the initial speed of the entire vehicle and to convert this parameter into the corresponding angular speeds via the simple relation[2]

$$\omega = \frac{v}{r} \tag{16.1}$$

Nevertheless, the specification of the initial speed is important, since depending on the implementation of the speed controller, attempts are made to control from the current (initial speed) to the target speed. If the two are too far apart, this can lead to unwanted acceleration or deceleration phases. For this reason, the speed curve should always be observed during the first tests.

16.1.2 Open-Loop Maneuvers

The open-loop control for the speed of the vehicle usually specifies only one drive or brake torque over time or over distance (Figs. 13.2 and 14.2). It is therefore primarily suitable for simulating acceleration or deceleration situations. Adherence to a set speed cannot be checked and cannot be guaranteed. The same drive torque does not have the same effect with different load conditions or road courses.

Nevertheless, this type of speed setting can be a good choice. If a defined maneuver is to be implemented without great effort and no suitable controller is yet available or if the exact maintenance of a speed is not really important.

When specifying drive torques, care should be taken to ensure that the specification corresponds to realistic values at the point of discharge in terms of both its magnitude and its time response (gradient). The slip values of the driven tires can provide an indication here, since increased slip values will occur with too large torques or torque steps (slippage of the tires).

Depending on the modeling depth, the drive torques are applied directly on the wheels, axle differential or center differential. If the engine and gearbox are modeled at least as vibrating masses, then the classical NEWTON's *actio et reactio* will act. This means that the delivered torque must be supported on both sides (input and output side).

---

[2]Depending on the simulation environment, all bodies may also have to be parameterized with an explicit initial position and velocity. Some programs use the knowledge that the bodies are connected to each other and therefore (depending on the binding) must have the same speed.

16.1.3 Closed-Loop Maneuver

For many maneuvers it is necessary to maintain a constant speed or a given speed curve. This is not possible without control. In the real vehicle, this is the responsibility of the driver, who can do this more or less well. Test drivers on test tracks are of course much more constant than normal drivers on public roads. This function is comparable to an automatic cruise control. Here, however, it must be considered whether a speed reduction is only achieved by the drag torque of the engine or also by braking. Depending on the system used, this is solved differently.

The simplest conversion is a torque specification on the wheel proportional to the deviation from the target speed—a so-called P-controller.

$$\Delta v = v_{shall} - v_{is} \tag{16.2}$$

$$M = p \cdot \Delta v \tag{16.3}$$

The formulation in Eq. (16.3) does not distinguish between a drive torque and a brake torque. Is the instantaneous speed v_{is} greater than the target speed v_{shall} (vehicle is too fast), a negative speed difference Δv results, which, multiplied by a proportionality gain factor p, results in a brake torque M. If the difference is positive (the vehicle is too slow), the torque is also applied positively and accelerates the vehicle. The larger the factor p, the faster speed differences are corrected and the fluctuation range of the speed is small. Even small differences are weighted accordingly. This small fluctuation around the target speed cannot usually be observed in measurements of real vehicles, so that this value should be iteratively adapted to the behavior of the vehicles in this driving situation.

However, there is also an increased risk that excessive torques will be transmitted and this will cause the wheels of the drive axle to spin. Figure 16.2 shows how sensitive this parameter is for an acceleration maneuver. With a p-value of 590, the vehicle model used here achieves maximum power transmission with approx. 18% slip.[3] At a value of 593 the slip increases to 19% and the vehicle accelerates slightly worse. Already at a value of 594 the drive wheels spin, which is visible by a slip value of approx. 93%. The gain values only apply to the vehicle model used here and the driving situation. Ultimately, only a comparison with the acceleration behavior of the real vehicle can show how large the factor must be chosen. It is also advisable to limit the torque as shown in Eq. (16.4). Especially if you do not want to look at the course of the torque every time.

$$M = \min \left(\max \left(p \cdot \Delta v; M_{max,brake} \right); M_{max,accel} \right) \tag{16.4}$$

---

[3]Slip definition according to Eq. (12.3).

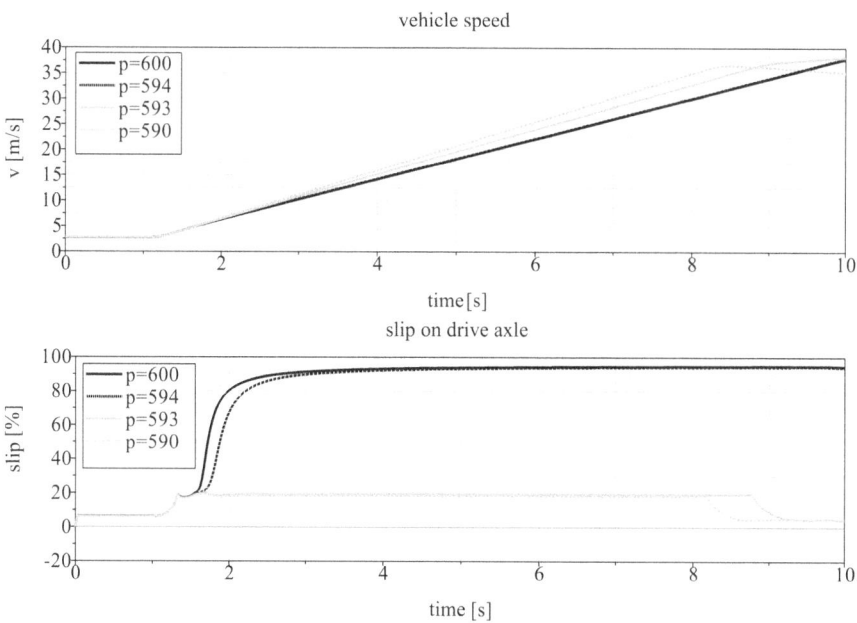

Fig. 16.2 Influence of the gain factor on speed control

This is a suitable approach for maintaining a speed. If acceleration or braking maneuvers are to be simulated, this modeling is often inadequate, as the presented example has shown.

At this point we would like to point out that it may be advisable to compare the recorded vehicle speed with the simulated one in the case of a simulation and measurement comparison for some (very speed-sensitive) maneuvers. However, it must be noted that the way in which the vehicle speed is measured plays a significant role. It must always be distinguished whether the speed is measured with slip or not. All speeds, which are determined from the wheel speed, are subject to slip and therefore usually too fast in the case of drive. If the vehicle speed is measured directly to the road surface (e.g. by radar or optical means), it is directly comparable with the simulated speed, which is normally determined in relation to the inertial system.

If braking and acceleration maneuvers are to be simulated, the driver's request after deceleration or acceleration must be passed on to a model of the brake system (Chap. 14) or the drive train (Chap. 13) and the corresponding torque must then be calculated and applied there. Only in this way can the time behavior of the torque curve be adequately mapped.

16.2   Steer Control

Similar to speed control, steer control must ensure that a vehicle would never drive straight ahead without steer angle control. So it is not enough not to steer in order to drive straight ahead. The road inclination, asymmetries in the weight distribution or the drive torques, which cannot be transmitted symmetrically due to different coefficients of friction or wheel loads, ensure that the vehicle runs to one side, real and virtual. Also here the hint at the beginning to check the driven trajectory, this can usually already be done with the help of the animation of the calculation results.

16.2.1 Open-Loop Maneuver

In real drive tests, there are some steering maneuvers that are open-loop. The steering wheel angle is changed in a given way and the vehicle reaction is observed. The most prominent example is certainly the step steer according to [ISO7401]. Here, at a vehicle speed of 80 km/h with a steering angle speed of 200–500°/s, the vehicle is steered in such a way that a lateral acceleration of $a_y = 4$ m/s^2 is reached. The development and the overshoot behavior of the yaw rate are evaluated. The implementation of this maneuver in the simulation is based on the same specifications as in the drive test and can be taken from the standard or the test regulations.

16.2.2 Closed-Loop Maneuver

Steer control is becoming more complex. Here, it is not enough to integrate a P-controller that immediately compensates for any deviation. That way, a nervous driving style is created, which can very quickly lead to unstable driving behavior if the deviations are too large.

What happens to us real drivers? According to [Dong78], the steering control process consists of two levels. The anticipatory control (open-loop) and the compensatory control (closed-loop) (Fig. 16.3).

This subdivision can be clearly explained using the example of a novice driver and an experienced driver. The novice driver initially does not have sufficient experience of which steering wheel angle to turn for the given vehicle so that the vehicle remains on the desired trajectory at the current vehicle speed, road condition and curvature. As a result of this lack of experience, the result must be found iteratively, since every time the actual trajectory deviates from the target trajectory, a steer angle correction must be carried out. The experienced driver intuitively selects a steer angle that, under known road conditions, generally leads to the desired lane without major changes in the steer angle. Only minor corrections are necessary. This is not a conscious process. Nobody calculates the required angle before passing through the curve. Most drivers will not be able to quantify this angle

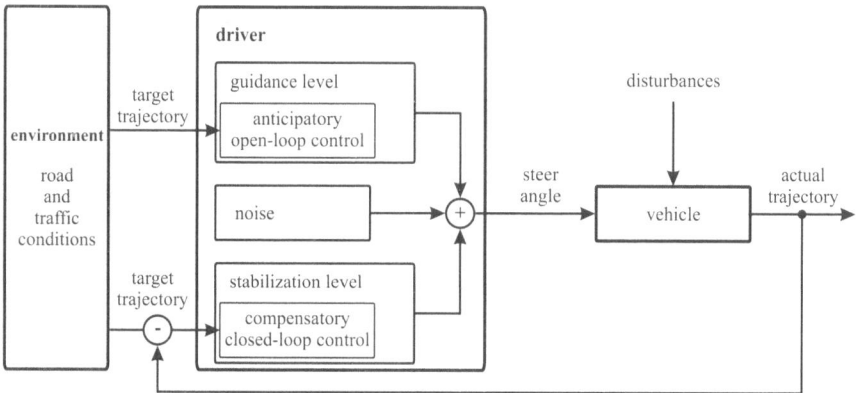

Fig. 16.3 Two-level steer control model

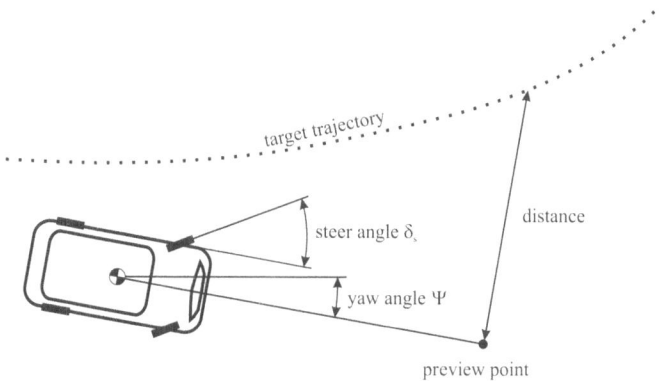

Fig. 16.4 Drawbar model

at all. It is simply the experience that enables him to do this backmarked. How can this be represented in the simulation?

The most common model in the literature is the so-called drawbar model (Fig. 16.4), which works with a preview point. Because of this preview point, the deviation from the actual and target trajectories is not compared directly in front of the vehicle, but at a certain distance. The distance between the point and the vehicle reference system may vary depending on the maneuver and driving speed. Here, too, one can observe parallels to the example of beginners and experienced drivers. Driving beginners often do not dare to look too far in front of the vehicle in the initial phase of their driving skills and for this reason also have to correct more frequently. The further you look ahead, the earlier you can estimate the course of the route.

On a winding route, the distance between the preview point and the vehicle will be less than on a more straight or slightly curved route. If the same route is travelled at different vehicle speeds, it has proven to be useful to select the preview point depending on the speed. The higher the speed, the further away the point should be. That way, deviations from the nominal course can still be compensated by minor steering angle corrections.

Section 12.5 has already described how road information can be prepared for ride comfort and load data prediction so that the tire can, for example, run over a digitized real road profile. In addition to this information on profile height and coefficient of friction, an important question is on which lane this road should be crossed. For this a trajectory must be available, which makes this information available to the steer controller. If you vary this trajectory slightly, you can estimate the influence of the lane on the result of stochastic road profiles. Since in very few cases a real driver can reproduce his lane identically, this is a valuable piece of information, especially for a simulation-measurement comparison.

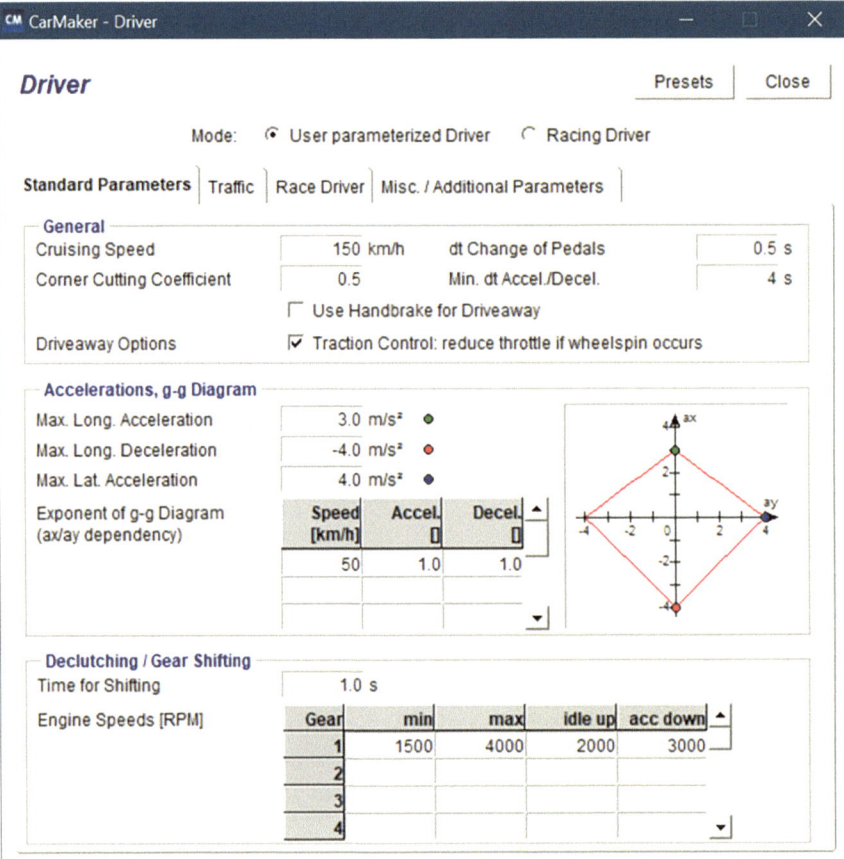

Fig. 16.5 Parameterization of the driver model using the example of the IPG-DRIVER

For vehicle dynamics simulations, the trajectory driven can have a major influence on lap times, especially in racing. Steer controls for this application allow to adjust the way curves are driven through.

16.3   Complex Driver Models

In addition to following a course and maintaining a speed profile, models can also represent different types of drivers. That way, the behavior of normal drivers, sporty drivers, test drivers and even racing drivers can be simulated.

If different driver types are to be simulated, this is usually controlled by exploiting the longitudinal and transverse force potential. With the help of a g-g diagram, the driver can be controlled depending on the direction of travel to be more courageous or even more cautious (Fig. 16.5).

In racing mode, the main focus is on finding the optimal trajectory and braking points by utilizing maximum longitudinal and lateral forces. This is partly due to the *training* of the driver model. In this way a lap time optimization can be carried out. In contrast to the previously described procedure, a target trajectory and a target velocity curve are no longer specified, but a lane in which the model is to find the optimum trajectory and the appropriate velocity profile. As this depends on the road conditions (coefficient of friction, course of the road) and the vehicle, this is determined iteratively.

The structure of such driver models is described for example in [Fisc11] or [IPG05].

References

[Dong78]   Donges, E.: Ein regelungstechnisches Zwei-Ebenen-Modell des menschlichen Lenkverhaltens im Kraftfahrzeug, *Zeitschrift für Verkehrssicherheit*, Bonn, 1978

[Fisc11]   Fischer, R. et al: Fahrermodell zur virtuellen Regelsystementwicklung, *ATZ 12/2011*, 946-949, 2011

[IPG05]   IPG, *Virtual Driver Model*, Vehicle Dynamics International, 2005

[ISO7401] DIN ISO 7401: *Testverfahren für querdynamisches Übertragungsverhalten*, 1989

[Mayr01]   Mayr, R.: *Regelungsstrategien für die automatische Fahrzeugführung*, Springer Verlag, Berlin, 2001

The Vehicle Model as a Controlled System 17

Most of the components described above are available both in the passive form and as an integral part of an active chassis system (Fig. 17.1). After the simulation of chassis and complete vehicle models in the 1990s was promoted and also demanded primarily by the development of vehicle dynamics control systems such as ABS, TCS or ESC, today the many assistance systems are the drivers for simulation. Irrespective of whether they are designed as comfort or safety systems, they require fast and reliable models and simulation processes in order to design and secure the wide variety of vehicle variants.

17.1 Development of Control Systems

An essential basis for the design of complex control systems is the sufficient mapping of the controlled system. Only a vehicle model whose performance has been adapted can ensure the subsequent use of the control algorithms in the real vehicle. Simple linear models may still be the first choice for preliminary considerations and some sub-areas. They are fast and manageable in the truest sense of the word. Often, this procedure is not suitable for transfer to the vehicle. The possible interactions due to the control interventions within the chassis require a closer look. For example, the vibrations generated by the pulsation of an ABS control in the wheel suspension cannot be detected in a single-track model. Here, the interplay of brake control, the elastic connection of the wheel suspension, the brake hydraulics and the tire contact with the road must be modeled in more detail.

© Springer Fachmedien Wiesbaden GmbH, part of Springer Nature 2021   277
D. Adamski, *Simulation in Chassis Technology*,
https://doi.org/10.1007/978-3-658-30678-6_17

[Daimler AG]

Fig. 17.1 Hardware-in-the-loop test bench

17.1.1 Software-in-the-Loop

In the context of model-based development, it is particularly interesting for controller developers to test their controller model at an early stage in its later environment. The algorithms of the controller code are initially available in software form, since they can usually be used with corresponding tools such as MATLAB/SIMULINK or ASCET have been developed which have the function of code export (Fig. 17.2).

Frequently, especially in the early phase of controller development, only the actual controller function is mapped. Elements such as safety and monitoring routines are not yet integrated. If this controller code is now integrated into a system environment that represents the vehicle, this is referred to as software-in-the-loop simulation or SiL for short. If the monitoring of the time behavior is not yet implemented, the individual system models do not have to be real-time capable, so that the expensive hardware required for this can be dispensed with.

The SiL simulation can be implemented as part of a co-simulation with a vehicle or chassis model or with a so-called autobox in a real vehicle. Once the desired maturity level of the controller model has been reached, code export functions can be used to generate C code from the model, for example, which may then already have been prepared for the later target platform. More about this can be found for example in [Kock08].

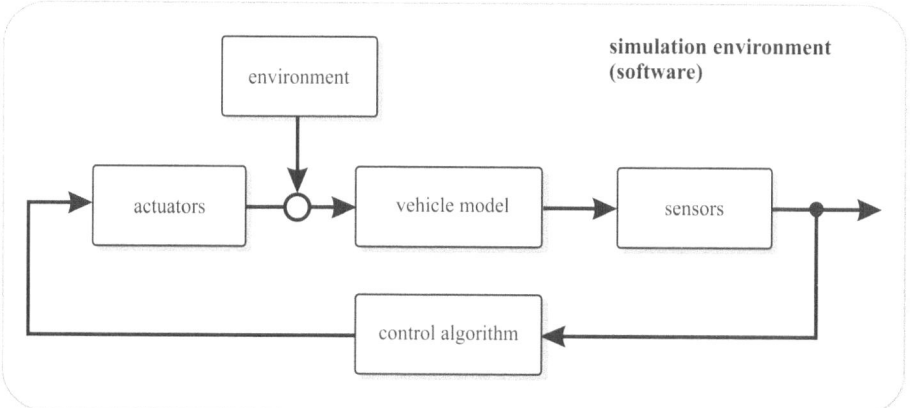

Fig. 17.2   Basic structure of software-in-the-loop

17.1.2 Hardware-in-the-Loop

17.1.2.1 General Conditions

A development strongly driven by the E/E architecture of the vehicle is hardware-in-the-loop simulation, or HiL for short. It is used to integrate components into a subsystem or overall system, which means that the component is not considered in isolation, but in interaction with other components or systems. The component is available as hardware and the rest of the vehicle is simulated. The target control unit is often integrated with the control algorithm as a component (Fig. 17.3). In the same principle, however, a part of the actuators or sensors can also be available as hardware.

The component is thus pretended to be installed in a real vehicle (Fig. 17.1). In [VDI2206] it is demanded: "The HIL simulation of dynamic systems by physical and mathematical models must take place in real time and under simulation of the physical loads".

Most real components that are examined using HiL simulation have the requirement that they must be operated in real time. What does that mean? Let us take as an example for the component a control unit for an active chassis system. This control unit receives new signals from a sensor in a fixed cycle, say, every 20 ms, and sends signals to an actuator in another cycle, say, every 50 ms. Security software monitors the proper reception and successful transmission. If a deviation is detected, an error is reported and under certain circumstances the desired functionality is switched off. From this requirement arises the constraint for the simulated environment of the control unit that it must be able to deliver and also accept information within the system's own cycle times. If the behavior of the vehicle is monitored within the safety software, then this must also be *punctually* otherwise implausibilities are detected and a corresponding error message is triggered.

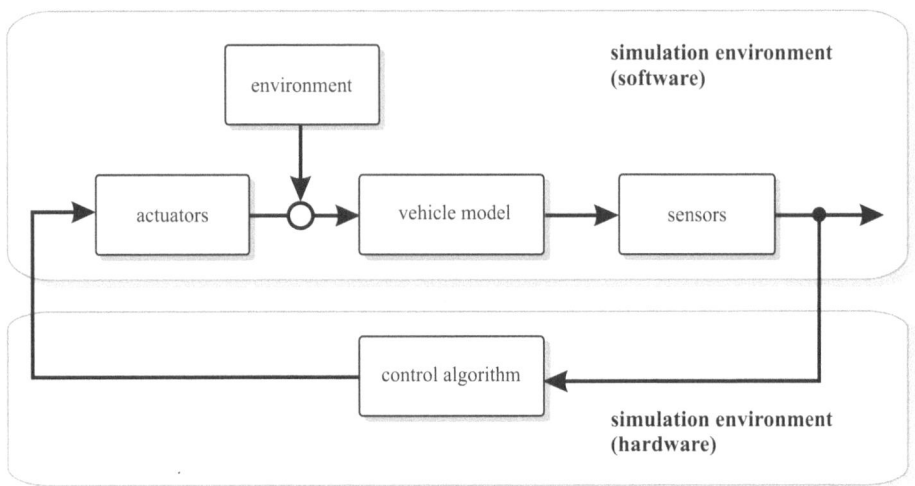

Fig. 17.3 Basic structure of hardware-in-the-loop

The term "real-time" therefore means that the simulated system can be calculated so quickly that the real component does not detect any delays. In a defined time window, for our example we assume 5 ms, so the system behavior for this time window must be guaranteed to have been calculated, that is, the required computing time must be below the 5 ms.

In summary, it can be said that the required computing time must always be less than the simulated time (Fig. 17.4). A reserve to be defined for the respective system behavior is necessary in order to be able to compensate for fluctuations in the calculation steps.

17.1.2.2 Hardware and Process Requirements

Frequently, the ECUs installed as hardware also need information that is not directly related to the planned tests. Nevertheless, they must be made available, since otherwise the function of the control unit is impaired, since a partial or even the entire function may have been switched off by corresponding error messages. In order for all signals or information to be available, they must be generated artificially. Even with these signals, you have to take into account the correct values, their format and their changes during operation. Otherwise problem-free operation cannot be guaranteed. If the control unit receives its information, for example via a communication bus (CAN, FlexRay, etc.), a so-called remaining bus simulation must be carried out. In addition to the user information, it also contains the necessary additional information. Which these are, depends on the control unit and must be checked on a case-by-case basis. The physical loads mentioned above mainly affect the electrical side, i.e. the control unit must have the correct current, voltage and resistance.

Before the simulation, the built-in control unit must be checked for possible error entries that would prevent proper operation. For this reason, it is necessary that the associated

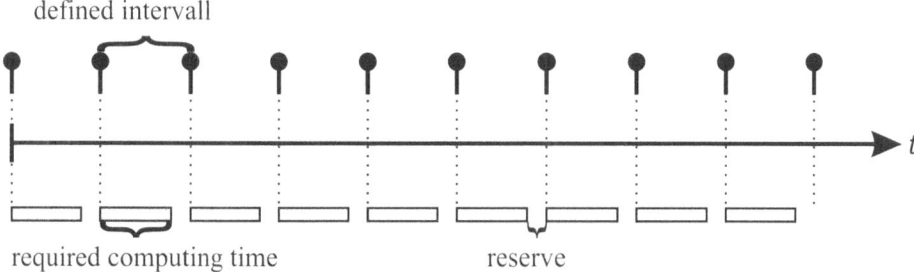

defined intervall

required computing time   reserve

Fig. 17.4 Real-time condition

diagnostic software can also be used within the framework of HiL simulation (preferably automated).

Overall, the initialization routines must be set up in such a way that the control unit equals a vehicle restart. To start a, for offline simulation *more normal,* condition already with a start speed of $v_0 > 0$ is not possible for most control devices, since they (rightly) consider this implausible and usually exit with a corresponding error message.

The necessary computer architecture together with a real-time operating system can often be obtained from commercial providers. However, it is always necessary to adapt the system to your own vehicle and chassis structure, which should not be underestimated, and the desired test scenarios must also be created according to your own requirements.

In contrast to offline simulation, 100% reproducibility is no longer guaranteed, depending on the hardware used. Additional factors responsible for the result are signal run times, processor clocks, bus load, temperature and many more things. The consequence of this is that, especially in automated operation, not hard limit values but tolerance bands for the expected results should be taken into account.

17.1.2.3 Functional Test
The functions which are to be covered by the system to be tested are described in a specification and often assigned setpoints and limits. In order for compliance with the specification to be checked, each point in the specifications must be approached and compared with the corresponding values. These checks should be automated and should produce a log documenting compliance (or malfunction).

17.1.2.4 Error Simulation
The error simulation is an extension of the functional test described above. Here, too, it is conveniently automated to check how the entire system reacts to specified error patterns. A distinction must be made between different error classes. For example, sensor signals can fail, be disturbed (EMC), amplified or shifted (offset). Each specified error is stored in the error memory of the control unit by an error message and can be read out by the diagnostics. Whether one has really considered all possible error patterns in advance and then also checked, can only be guaranteed by exact system knowledge and the appropriate

experience. The number of tests increases over time, as gaps are recognized retrospectively and new findings can be incorporated.

17.2   Sensors

Sensors in the vehicle are usually discrete sensors, which means that they transmit the measurement signal only at certain, mostly equidistant points in time. On the other hand, the continuous original signal is discretized, i.e. represented in a finite stepped resolution (see also Sect. 2.2.2). If the controller which is supplied by this sensor expects a time and/or value discrete signal, the calculated signal must first be converted. Each measured quantity of the simulation model is of course also a discrete quantity, since it is only queried at certain times (depending on the time step of the integration procedure or the sampling rate for this quantity) and since it is resolved in a finite number format (e.g. with double precision, i.e. in an 8-byte representation).

If the controller is supplied with the quasi-continuous signal, it will be practically *oversupplied*. This means that it gets a better signal quality than the real controller. In this way, the control quality may be better and the vehicle model may behave more favorably than the real vehicle. For this reason, you should make sure that the controller receives a signal that is discrete both in time and in value, just as it is in reality. Since the integration step size or the sampling of the calculation will not always harmonize with the sampling rate of the sensor or the controller, a time management must be integrated which can interpolate intermediate values if necessary (see also Sect. 4.5).

17.3   Actuators

Real actuators have a time behavior that must be taken into account if one does not want to achieve too optimistic results. A valve that can open or close immediately without delay promises a good control result—only one supplier for such a valve will probably be sought in vain. Here, for serious investigations, it is necessary in any case to get measurements or data of the component manufacturer, which time behavior is to be expected.[1] Depending on the type, our example valve still has a differentiable time response for the opening and closing process.

If the actuator model is used in a control loop of a HiL or SiL simulation, the following must be observed: If the controller expects a certain time behavior, this must also be taken into account in the model, as otherwise the controller may be switched off incorrectly, as it considers the behavior of the actuator to be implausible. More details can be found in Sect. 17.1.

---

[1]Which may depend on the temperature, the state of the electrical system and many other influences.

MBS programs are primarily designed to represent mechanical components. However, if the active system is strongly influenced by electrics, pneumatics or hydraulics, it can be useful to integrate third programs in order to be able to present these disciplines with the necessary detailing (and with the necessary comfort). Please refer to Sect. 5.2.3.5 for further details. If the description can be limited to a few differential equations, this can usually be done without problems in the MBS program itself with a little experience.

References

[Kock08]   KOCK, P. ET AL: Komponententest von modellbasierten Funktionen im MIL/SIL und PIL Test mit Anwendungen von formalen Testmethoden, in *Simulation und Test in der Funktions- und Softwareentwicklung für die Automobilelektronik II*, Expert Verlag, Renningen, 361-377, 2008

[VDI2206]   VDI-Richtlinie 2206: *Entwicklungsmethodik für mechatronische Systeme*. Verein Deutscher Ingenieure, Duesseldorf, 2004

Index

A

ADAMS-BASHFORTH method, 61
ADAMS-MOULTON method, 61
Adhesion, 213
Alias effect, 22
Analog system, 22
Anti-lock brake system (ABS), 247
Anti-roll bar, 171
Archiving system, 108
ASCII format, 76
Auto scaling, 79

B

Benchmark, 74
Bending line, 166
Binary form, 75
Black box system, 24
Brake slip, 216

C

Causality, 16
Cause, 16
Characteristic curve, 95
Characteristic map, 95
Cleat, 228
Closed-loop maneuver, 267
Coil spring, 162
Communication effort, 86
Component tolerances, 40
Concentrated mass, 261
Concept of systems, 13
Consistency, 116

Consistent data and models, 104
Constraint Modes, 259
Contact patch, 211
Continuous system, 22
Convergence, 53
CRAIG-BAMPTON, 259
Crosswind, 263
Cubic spline, 64
Curved Regular Grid (CRG), 232

D

Database, 104
Deduction, 31
Degree of confidence, 28
Descriptive name, 92
Design parameter, 90
Deterministic system, 109
Development release, 104
Deviation moments, 103
Differential Algebraic Equations (DAE), 77
Digital system, 22
Discrete system, 22
Discretization of the time behavior, 22
Drive slip, 216
Dynamic analysis, 82
Dynamic stiffness, 142
Dynamic system, 21

E

Effect, 16
Electric power steering (EPS), 203
External drum test rig, 228

F

Fast Fourier Transformation (FFT), 79
Flat track test rig, 228
Force-stroke characteristic curve, 162
Freeway Hop, 235
Full vehicle validation, 42

G

GUI license, 107
GUYAN, 259

H

HEUN, 59
Highway Safety Research Institute (HSRI), 222
Homogeneity 261
Initial speed, 269
Instantaneous pole, 39
Internal drum test rig, 228
Isolated degree of freedom, 126

K

Kernel license, 107
Kinematic analysis, 82

L

LC circuit, 14
Level control system, 177
License, 107
Linearization, 19 82
Local error, 50
Log file, 107
Longitudinal slip, 217
Loss angle, 142
Lumped masses, 260

M

Mass distribution, 261
Mass moments of inertia, 102
Master brake cylinder, 249
Misuse, 116 224
Modal damping, 259
Multi-link suspension, 128
μ-split braking, 213

N

Naming convention, 92
Naming of the parameters, 91
Nomenclature, 91
Nominal component, 40
Normal Modes, 259

O

OpenCRG, 232
Open-loop maneuver, 267
Ordinary Differential Equations (ODE), 77

P

Postprocessor, 75
Power spectral density (PSD), 79
PQR-method, 140
Predictor-corrector technique, 58
Prefix, 93
Preprocessor, 75
Preview point, 273
Principal axis of inertia, 102
Process milestone, 105

R

RBE2, 258
RBE3, 258
Real time, 279
Reduction method, 259
Repeatability of measurements, 40
Rim, 209

S

Sampling rate, 22
Sampling theorem, 22
Scattering, 40
Secondary spring rate, 137
Sensor position, 39
Separation, 15
Settling time, 121
Simulator, 116
Single-mass oscillator, 14
Single-source principle, 104
Single-tube damper, 183
SI units, 94

Solver, 75
Speed controller, 269
Spring ratio, 179
Spring stiffness, 162
Sprung mass, 161
Stability, 53
Standard Tyre Interface (STI), 218
State vector, 121
Static analysis, 82
Static deformation, 257
Static system, 21
STEINER's theorem, 103
Step steer, 272
Stochastic system, 109
Substitute system, 24
Suffix, 94
Superposition, 19
System boundary, 14
System environment, 15
System identification, 24
System with memory, 21

T
Test case, 34
Time-continuous system, 22
Time-discrete system, 22

Tire, 209
Tire Model Performance Test (TMPT), 227
Toegepast Natuurwetenschappelijk Onderzoek
    (TNO), 220
Top-down method, 30
Top mount, 154
Total mass, 261
Track length, 122
Triangle inequality, 102
Twin-tube damper, 183
Tyre Data Exchange Format (TYDEX), 218

U
University of Michigan Transportation Research
    Institute (UMTRI), 222
Unsprung mass, 161

V
Vehicle reference system, 100
Vehicle-specific parameterization, 31
Version control, 104
Viscous damping, 183

W
White-box system, 24